A STANDARD FOR
ENTERPRISE
PROJECT
MANAGEMENT

ESI International Project Management Series

Series Editor

J. LeRoy Ward, Executive Vice President

ESI International

Arlington, Virginia

Practical Guide to Project Planning
Ricardo Viana Vargas
1-4200-4504-0

The Complete Project Management Office Handbook, Second Edition
Gerard M. Hill
1-4200-4680-2

Determining Project Requirements
Hans Jonasson
1-4200-4502-4

A Standard for Enterprise Project Management
Michael S. Zambruski
1-4200-7245-5

Other ESI International Titles Available from Auerbach Publications, Taylor & Francis Group

PMP® Challenge! Fourth Edition
J. LeRoy Ward and Ginger Levin
ISBN: 1-8903-6740-0

PMP® Exam: Practice Test and Study Guide, Seventh Edition
J. LeRoy Ward
ISBN: 1-8903-6741-9

The Project Management Drill Book: A Self-Study Guide
Carl L. Pritchard
ISBN: 1-8903-6734-6

Project Management Terms: A Working Glossary, Second Edition
J. LeRoy Ward
ISBN: 1-8903-6725-7

Project Management Tools CD, Version 4.3
ESI International
ISBN: 1-8903-6736-2

Risk Management: Concepts and Guidance, Third Edition
Carl L. Pritchard
ISBN: 1-8903-6739-7

A STANDARD FOR ENTERPRISE PROJECT MANAGEMENT

MICHAEL S. ZAMBRUSKI

CRC Press
Taylor & Francis Group
Boca Raton   London   New York

CRC Press is an imprint of the
Taylor & Francis Group, an **informa** business

AN AUERBACH BOOK

Auerbach Publications
Taylor & Francis Group
6000 Broken Sound Parkway NW, Suite 300
Boca Raton, FL 33487-2742

© 2009 by Taylor & Francis Group, LLC
Auerbach is an imprint of Taylor & Francis Group, an Informa business

No claim to original U.S. Government works
Printed in the United States of America on acid-free paper
10 9 8 7 6 5 4 3 2 1

International Standard Book Number-13: 978-1-4200-7245-7 (Softcover)

Library of Congress Cataloging-in-Publication Data

Zambruski, Michael S.
A standard for enterprise project management / Michael S. Zambruski.
p. cm. -- (ESI international project management series ; 4)
Includes bibliographical references and index.
ISBN 978-1-4200-7245-7 (alk. paper)
1. Project management--Standards. 2. Project management--Forms. I. Title.

HD69.P75Z36 2008
658.4'04--dc22
2008001430

Visit the Taylor & Francis Web site at
http://www.taylorandfrancis.com

and the Auerbach Web site at
http://www.auerbach-publications.com

Contents

Analysis and Planning

Execution and Control

Closure

List of Figures

Preface

Project management is about turning ideas into results. Unfortunately, it is commonly viewed in isolation from the other business disciplines that form the context needed for its success—namely, strategic planning and requirements analysis before the project, and operationalization after the project. As a result, uncertainty or confusion about the role of project management all too often arises, leading to questions such as these:

- What is the difference between the business vision, mission, and goals, and what do they have to do with projects?
- How does the business requirements document (BRD) differ from the project charter?
- Why do we need a project charter and a BRD?
- Why create a statement of work if you already have the BRD?
- Do we really need this much process in order to get a project going?

A Standard for Enterprise Project Management explains each of the basic elements needed for project success and integrates them into a balanced life-cycle continuum. It also supplies an inventory of practical policies, procedures, techniques, and templates for immediate use. The result is a handbook for getting the work done fast, smart, and right.

There are three components to the book. The first is the main body of text, which provides a description of logical project phases and associated documents, beginning with authorization and initiation, followed by analysis and planning, then execution and control, and finally closure. Each phase contains both an explanation and an illustration of what can be done to optimize success.

Throughout the main text are references to dozens of appendices found at the end of the book. They constitute the second and largest component, that is, blank and completed templates suggested for use. Each of these tools contains details on how to apply them, with emphasis on balancing the benefits of standardization with the need for flexibility.

The third component is the CD, which holds a full-color version of the base document with all the figures and appendices. The appendices are included as embedded files displayed as icons within the main text file. Double-clicking on an icon allows the embedded file to open for use. In this way all of the blank templates as well as the completed samples are instantly available and completely portable. In order to open all of these files, it is necessary to have Adobe® Reader as well as the following Microsoft® applications: Word, Excel, Visio, and Project.

At the end of the CD are four bonus items. Bonus 1 is a Quick Start with Project 2003. This is a one-page tutorial with three pages of screen prints designed to quickly generate readable and concise project plans. Bonus 2 is a Complex Project Readiness Grid. It is a matrix suggesting how to manage intricate interrelationships in a project or program environment. Bonus 3 is a Project Management Competency Development grid, which outlines a program for developing key skills among project managers within an organization. Bonus 4 is an example of Traceability in Business Analysis and Project Management, which shows a chain-of-custody relationship up and down the requirements-solutions continuum.

The best way to implement the concepts, processes, and tools in *A Standard for Enterprise Project Management* is to adopt them as the starting point for structured yet adaptable models of project success within an organization—from idea inception all the way to post-implementation production, and each step in between.

One note regarding the appendices: they are organized in a proposed numerical order that corresponds to standard project phases. However, this does not mean that every project must have every appendix in the exact sequence shown. Factors such as the project size, complexity, risk, duration, time sensitivity, and association with other initiatives will ultimately determine which tools are needed and when. Accordingly, discussions in the body of the book focus on the various appendices in terms of their relative importance or relationship to each other rather than their simple linear succession. As a result, references to the numbered appendices do not always occur within the main text in the exact order as shown in the table of contents.

About the Author

Michael S. Zambruski has been providing professional project management, business analysis, and training for a wide variety of multibillion-dollar international firms; small, fast-growing companies; and entrepreneurs for more than 25 years. His diverse background covers both the service and product sectors, with industry experience in telecommunications, information technology, health care, higher education, environmental services, consumer goods and services, advertising, banking, real estate, and aerospace, as well as in the federal government—both civilian and military.

His assignments have involved service and product management, process redesign, business development, engineering, manufacturing, quality control, product distribution, strategic marketing, and technology integration. His achievements include organizing multimillion-dollar projects, expanding market performance, designing financial decision models, creating new service concepts, leading crisis-reaction teams, and building project management offices at diverse organizations that include Unisys, UMass Memorial Medical Center, Yale University, CIGNA, Lucent, SBC, and Boeing. His first book, *The Business Analyzer & Planner* (AMACOM, 1999), presented a unique seven-phase methodology for understanding the fundamental issues behind problems and opportunities, and then mapping out alternatives for optimal results. His articles published by ESI International include "Organizing Structure in the Midst of Chaos" (2005), "Establishing Clear Project Management Guidelines" (2006), and "The Portability of Project Management" (2007).

Mr. Zambruski has taught business courses for Quinnipiac University, the University of New Haven, the University of Phoenix Online, and Boston University's Project Management program. He holds an M.B.A. from Southern Illinois University and a B.A./B.S. from Georgetown University. He is certified as a project management professional (PMP®) by the Project Management Institute (PMI®), of which he is a member and holds the Advanced Master's Certificate in project management from George Washington University. His e-mail address is Michael.Zambruski@snet.net.

About the Author

Chapter 1

Introduction

This handbook describes policies, procedures, techniques, and tools for the uniform management of projects throughout an organization. They combine standardization with responsive flexibility and best practices to achieve on-budget, on-schedule performance while carefully managing scope, quality, and risk for all projects regardless of size or complexity.

Figure 1.1 outlines the overall context of the processes, key documents, and activities specified in this handbook. It portrays project management as the logical sequence of how work should progress from the idea stage in the Business Domain to the implementation stage in the Operations Domain, with a clear indication at each step of what the deliverables are.

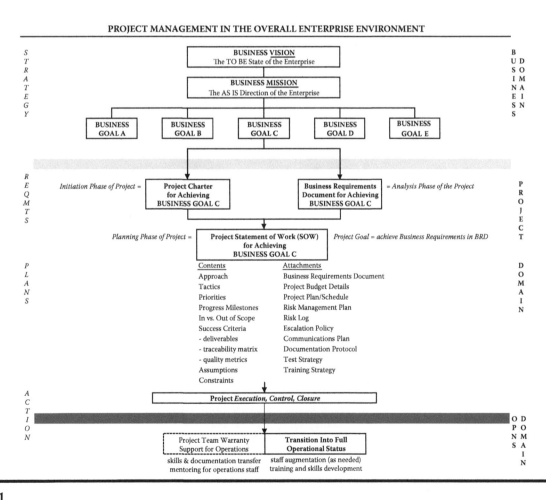

Figure 1.1

By portraying the Project Domain in this overall context, attention is drawn to those activities that must precede and follow a project in order for it to be considered a true success. In the Business Domain there must be a vision (or TO BE state) of the business and an associated mission (or AS IS state), together with the particular goals that support the vision and mission. These constitute the direction of the enterprise. In the Project Domain is where the more abstract elements of vision, mission, and goals evolve into concrete work delineated in the Charter, Business Requirements Document, and Statement of Work. Finally, in the Operations Domain the results of the project are implemented into day-to-day activities and thus represent the true improvements to the enterprise.

Chapter 2

Project Authorization and Initiation

2.1 Document Workflow

The appropriate amount of management and documentation for a project depends on many factors, including the project's size, duration, budget, complexity, and risk. The project assessment form in Appendix 2 can be used to help evaluate these factors and determine the level of project management needed for a particular initiative. Appendix 3 summarizes alternate paths for documenting various types of projects including those with a focus on information technology services.

2.2 Charter

Projects are authorized by means of a charter, which describes key high-level information—including *what* is to be done, a general timeframe for its completion, a summary of the budgetary resources needed and available, and key stakeholders. Once the charter is approved, as evidenced by the signatures at the end of the document, the project manager is authorized to begin work. Figure 2.1 shows a standard charter format. A charter template and completed sample are available as Appendices 4 and 5.

As a general policy, a copy of the completed charter should be forwarded to the internal audit and quality departments for determination as to whether an auditor and/or quality professional will be assigned to the project.

PROJECT CHARTER

Date:	Initiated by:
Project Summary Details	
Project Name:	
Project ID:	
Project Priority:	
Customer Name:	
Project Start Date:	
Planned Project End Date:	
Approved Budget:	*<<initial estimates are acceptable at this stage, to be refined as business analysis is completed>>*
Project Staffing Level (Total Person Months):	*<<initial estimates are acceptable at this stage, to be refined as business analysis is completed>>*
Project Personnel	
Project Sponsor(s):	
Business Owner(s):	
Project Manager:	
Other Key Personnel:	*<<typically included here are managers or key subject matter experts who will be instrumental to the success of the project>>*

Scope and Objectives:

<< The project scope and objectives are presented here at a high (executive) level. If a separate document contains that information, it can be embedded here.>>

Organizational Relationships (Roles and Responsibilities):

<<All organizations that will either contribute to this project or be impacted by it should be listed, together with the title of the person who will represent that organization for this project. This ensures that the organizational breadth of this project is clarified at the very beginning.>>

Key Dates or Milestones:

<<Very high level milestone dates are listed, to establish the general timeframe for the project.>>

Approvals:

Date: Signature: *<<electronic approval is acceptable if a dated email is referenced>>*

Date: Signature:

Figure 2.1

Chapter 3

Project Analysis and Planning

Once the charter is completed and initial funding is confirmed, the project formally begins and the documents around it become the blueprints for success. For modest initiatives, it may be sufficient to use condensed documentation such as the project summary template found in Appendix 6, which concisely presents all of the salient information on the project in a very condensed form. However, more sizeable projects are optimally served by more comprehensive efforts and records, beginning with the Business Requirements Document described next.

3.1 Business Requirements Document

The Business Requirements Document (BRD) specifies the concrete, measurable business improvements that are needed in order to achieve the high-level vision, mission, and goals of the sponsoring organization. It clarifies *why* the project is necessary and becomes a key reference for developing and implementing project deliverables. It is typically prepared by a business analyst who is assigned to the project team and who works closely with the project manager to identify, record, and validate the business requirements. Figure 3.1 shows the outline of a comprehensive BRD. A full template and a sample completed version are found in Appendix 8 and Appendix 9.

Business Requirements Document (BRD)

<< Project Name >>

Date Prepared (or Updated)	
Prepared by:	
Project Sponsor:	
Project Customer:	
Project Manager:	
Business Analyst(s):	
Project ID:	

TABLE OF CONTENTS

Figure 3.1

For large, complex, or lengthy projects, the preparation of a BRD normally involves extensive information collection, synthesis, documentation, and validation. These activities therefore benefit from careful planning. The Requirements Work Plan (RWP) shown in Figure 3.2 is used for this purpose. A template of the RWP is included as Appendix 7.

Business Requirements Work Plan (RWP)

<< Project Name >>

Date Prepared (or Updated)	
Prepared by:	
Project Sponsor:	
Project Customer:	
Project Manager:	
Business Analyst:	
Project ID:	

1. **Requirements Work Plan (RWP) Objectives**
2. **Deliverables**
3. **Scope**
4. **Participating Stakeholders**
5. **Resource Requirements**
6. **Assumptions**
7. **Dependencies**
8. **Constraints**
9. **Attachments**

   List and attach appropriate supporting and background documents – for example:

   - the project charter
   - the RWP analysis team roster
   - the RWP budget
   - a work breakdown structure (WBS) showing RWP activities
   - any other items that clarify the efforts to be undertaken to produce the BRD.

10. **Approvals**

Business Analyst	Date
Project Manager	Date

Figure 3.2

3.2 Statement of Work

The Statement of Work (SOW) specifies *how* the business requirements will be achieved and includes the overall project approach and tactics, a detailed timeframe with key milestones, funding details, success criteria, assumptions, constraints, and traceability to specific business requirements documented in the BRD. Essential to the SOW is a clear declaration of all activity that is in scope as well as out of scope. Figure 3.3 contains a standard format. A full SOW template and completed sample are found in Appendices 10 and 11.

STATEMENT OF WORK

<Project Name: _____>

<Project Period: mm/dd/yy – mm/dd/yy>

<Project Manager>
<contact information>

Table of Contents

Version History

1. PROJECT DESCRIPTION
   a) Goal
      i) strategic goal(s) and objective(s) supported by this project
      ii) significant risks
   b) Approach
      i)   normal application of Standard for Enterprise Project Management
      ii) crisis application of Standard for Enterprise Project Management
      iii) application of prior lessons learned
   c) Tactics
   d) Priorities
   e) Milestones
   f) Out of Scope

2. PROJECT TEAM

3. SUCCESS CRITERIA
   a) Key Deliverables
   b) Traceability to Business Requirements Document (BRD)
   c) Quality Metrics

4. ASSUMPTIONS

5. CONSTRAINTS

6. CHANGE CONTROL PROCESS

7. APPROVALS

ATTACHMENTS

A. Business Requirements Document
B. Project Budget Details
C. Project Plan
D. Issues/Risk Management Plan
E. Issues/Risk Log
F. Escalation Policy
G. Communications Plan
H. Documentation Protocol
I. Test Strategy
J. Training Strategy

Figure 3.3

Although the SOW serves as a principal reference document for all project efforts, changes are inevitable. Therefore, it is vital that a formal change management process becomes an integral part of managing the project. For this purpose a change request form template can be found in Appendix 20.

3.3 Project Team Roster

As early as possible a list of core team members, including any vendor staff, should be compiled. Contact information, area(s) of specialty and responsibility, and alternate representatives should be indicated for each person. This can be recorded in a stand-alone document or as part of the SOW. Figure 3.4 displays a standard project roster. Note that each member of the project team has an alternate identified, together with information on contacting those individuals as well as their administrative support staff. This is designed to ensure that no discipline goes unrepresented at key project meetings. If the primary is unable to attend, the alternate—who is totally aware of all relevant issues and actions—attends instead. A project roster template is located at the end of this handbook as Appendix 12.

PROJECT ROSTER

Name	Title	Department (or Firm)	Area of Specialty	Role/ responsibility	Phone contact	Alternate	Alternate's phone	Admin. Assistant
Czerny, Karl	Director, IT Projects	Info Tech	IT	telecom	123-456-7890	Sally	111-222-3333	123-456-7890
Effraim, Pete	Dir, Strat Space Plan	Planning	space plan	advisory	123-456-7890	Bob	111-222-3334	123-456-7890
Escrowel, Gene	Sr Dir, Critical Care	Critical Care	clinical	clinical sponsor	123-456-7890	Tom	111-222-3335	123-456-7890
Foley, Alex	COF	Finance	finance	financial support	123-456-7890	Vic	111-222-3336	123-456-7890
Fonce, Charlotte	Project Mgr	Engineering	equipment	equip, training	123-456-7890	Gina	111-222-3337	123-456-7890
Gentile, Gigi	Dir, Proj Mgt	Info Tech	applications		123-456-7890	Theresa	111-222-3338	123-456-7890
George, Tom	Chief of Nursing	Nursing	clinical	clinical coord.	123-456-7890	Bill	111-222-3339	123-456-7890
Letter, Christine	Dir, IT Opns	Operations	networks		123-456-7890	Mike	111-222-3340	123-456-7890
Orange, Phyllis	Dir. Fin Planning	Finance	finance	financial support	123-456-7890	Jem	111-222-3341	123-456-7890
Queen, Ellen	Dir, Quality	Quality	quality	gov't liaison	123-456-7890	Ackar	111-222-3342	123-456-7890
Silvan, Elliott	Proj Dir	Facilities	construction	constr proj mgt	123-456-7890	Tyrone	111-222-3343	123-456-7890
Sincol, Jerry	Vice-president	Cardio	clinical	clinical sponsor	123-456-7890	Sandie	111-222-3344	123-456-7890
Wuder, Sally	Sr. Cnsltnt, Labor Rel	Human Res	human res	labor, staffing	123-456-7890	Bob	111-222-3345	123-456-7890

Figure 3.4

3.4 Project Plan

The project plan document serves as the main control mechanism, both by specifying project phases and by decomposing these phases into specific tasks with associated timeframes, resources, dependencies, and deliverables. During project implementation, it also serves as a status tool by showing completion progress. It is typically included as *Attachment C* to the SOW and can be done in Microsoft Project or Excel, and possibly distributed as a document in Adobe pdf. Figure 3.5 is a segment of a project plan in Microsoft Project 2003. A standard template and samples of completed project plans can be found at the end of this book in Appendices 13.1, 13.2, 13.3 and 14.

PROJECT PLAN: Program Management Office Implementation

Michael S. Zambrusk
Tue 8/12/07

ID	% Complete	Task Name	Start	Finish	Duration	Predecessors	Resource
1	78%	1 INITIATE	Mon 7/16/07	Mon 8/6/07	16 days?		
2	84%	1.1 Develop Charter for PMO creation	Mon 7/16/07	Fri 7/27/07	10 days		Mike
6	50%	1.2 Obtain Charter signoffs	Mon 7/30/07	Mon 8/6/07	6 days?	2	
7	7%	2 PLAN	Tue 8/7/07	Mon 8/27/07	15 days?		Mike
8	35%	2.1 Reqmts Work Plan (RWP)	Tue 8/7/07	Thu 8/9/07	3 days?	6	
11	1%	2.2 Bus Rqmts Doc (BRD)	Thu 8/9/07	Mon 8/20/07	8 days?		
18	0%	2.3 Statement of Work (SOW)	Tue 8/21/07	Mon 8/27/07	5 days?	17	Mike, Bob
19	2%	3 EXECUTE	Mon 7/16/07	Wed 11/21/07	93 days?		
20	0%	3.1 Objective 1 - establish enterprise PMO	Tue 8/28/07	Mon 11/19/07	60 days		
24	2%	3.2 Objective 2 - Integrate Bus Analysis+Proj Mgt	Mon 7/16/07	Wed 11/21/07	93 days		
28	7%	3.3 Objective 3 - Implement PMO guidelines	Mon 7/16/07	Wed 9/5/07	38 days		
31	0%	3.4 Objective 4 - develop training curricula & sched	Tue 8/28/07	Fri 11/2/07	49 days?		
37	0%	4 CONTROL	Wed 10/3/07	Wed 12/19/07	56 days?		Mike, Team
38	0%	4.1 BA training	Mon 10/22/07	Fri 12/14/07	40 days	344	
39	0%	4.2 PM training	Mon 10/22/07	Fri 12/14/07	40 days	34	
40	0%	4.3 Project Mgmt Audits	Thu 11/22/07	Wed 12/19/07	20 days	27	
41	0%	4.4 Business Analysis Forums	Wed 10/3/07	Wed 12/5/07	46 days		
45	0%	4.5 Project Mgmt Forums	Fri 10/5/07	Fri 12/7/07	46 days		
49	0%	5 CLOSE	Thu 12/20/07	Wed 1/30/08	30 days	37	Team
50	0%	5.1 Operationalization	Thu 12/20/07	Wed 1/30/08	30 days		
51	0%	5.2 Lessons learned	Thu 12/20/07	Wed 12/26/07	5 days		

Figure 3.5

Chapter 4

Project Execution and Control

In order to optimize proper completion of approved project tasks, the following protocols should be defined and regularly followed throughout project implementation.

4.1 Issues and Risk Management

Identifying, recording, analyzing, and managing issues and risk are collaborative efforts of the project team and sponsor. They should begin as soon as the project is approved, but no later than commencement of project implementation. The issues/risk management plan and issues/risk log are typically included as *Attachments D* and *E*, respectively, in the SOW. Figure 4.1 is a simple issues/risk management plan. Appendix 15 offers a complete template.

Issues/Risk Management Plan

1. Any elements of conflict or uncertainty that can impact project success will be identified and proactively managed for optimal results. The primary mechanism for this process will be the **Issues/Risk Log** (normally Attachment E to the standard Statement of Work).

2. The Issues/Risk Log will provide the following information at a minimum for all OPEN items:
    a. organizational unit(s) affected
    b. unique number assigned to each item in the log
    c. full item description
    d. person identifying the item
    e. date entered into the log
    f. estimated <u>impact</u> on the project scope, schedule, budget, or deliverables
    g. estimated <u>likelihood</u> or probability of occurrence
    h. preliminary action needed, together with person assigned to manage the action
    i. ready status (red/yellow/green)
    j. regularly scheduled summaries of progress to date (e.g., weekly, monthly, etc.).

3. The Issues/Risk Log will provide the following information at a minimum for all CLOSED items:
    a. organizational unit(s) affected
    b. original number assigned to the item being closed
    c. full item description
    d. person who originally identified the item
    e. date originally entered into the OPEN log
    f. date closed
    g. person closing the item
    h. reason for closure.

4. The entire Issues/Risk Log will be updated at least once per week and will constitute a standing agenda item in all project team meetings, regardless of their frequency.

5. No-fault escalation to the next higher level of management will occur when any issues or risks remain open for longer than the maximum period allowable for the project, as specified in the **Escalation Policy** (normally Attachment F to the standard Statement of Work).

Figure 4.1

Figures 4.2a and 4.2b show the issues/risk log that accompanies the issues/risk management plan. Together they ensure consistent handling of risks and issues. An Issues/Risk Log template and completed sample are located in Appendices 16 and 17.

EMERGENCY COMMUNICATIONS PROJECT (Open risks)

Michael S. Zambruski
123-334-2438

ISSUES/RISK LOG

DEPT	RISK No.	RISK DESCRIPTION	IDENTIFIED BY	DATE ENTERED IN THIS LOG	IMPACT on successful survey	LIKELIHOOD of inclusion in survey	ACTION PLAN [Assignee]	READY STATUS	PROGRESS as of 11/21/2006
								R	less than 50% READY
								Y	50-75% READY
								G	greater than 75% READY
Police Precinct #1	1	**Processes** are not being followed	Officer Taylor	11/14/2006	High	High	To be covered in training push. Must track progress weekly. [Victoria]	Y	11-21-06 [Victoria] Written procedures are now available, and rehearsals are being planned.
St. Mary's	2	**Education** delivery	Chip Williams	11/21/2006	Med	Med	To be covered in training push. Must track progress weekly. [Sr. Pancratia]	Y	
Municipal Human Resources	3	Orientation and readiness of **new employees** (contract)	Donna Swanson	11/21/2006	Low	Low	Facilities and HR working on plan to prepare self-directed orientation for new hires. [Cicely]	Y	11/19/2006 [Karen] Ready by 12/15/06.

Figure 4.2a (Open risks)

EMERGENCY COMMUNICATIONS PROJECT (Closed risks)

ISSUES/RISK LOG

Michael S. Zambruski
123-334-2438

DEPT	RISK NUMBER	RISK DESCRIPTION	IDENTIFIED BY	DATE ENTERED IN THIS LOG	DATE CLOSED	CLOSED BY	REASON FOR CLOSURE [Decision-maker]
Cross-functional	4	**Infrastructure** support – turnaround time on contract and policy approvals is taking too long.	Mike Zambruski	12/10/2006	12/15/2006	Mike Zambruski	Joint subcommittee was created to accelerate infrastructure issues. [Victor]

Figure 4.2b (Closed risks)

4.2 Escalation

Especially with complex projects, a formal escalation policy is needed to ensure timely resolution of tasks, issues, and decisions which involve negotiable or debatable viewpoints. It is typically included as *Attachment F* to the SOW. Figure 4.3 is a sample; a template is available as Appendix 21.

ice, all items marked as RED in the Issues/Risk Log are subject to this policy.

tions will be documented at least via email to all involved/affected parties.   Such escalation

up the management hierarchy until a final decision is reached. Accordingly, each member of

e team must be made aware of his/her potential role in this process.

Escalation Policy is normally included as Attachment F to the standard Statement of Work.

y the responsibility of the Project Manager to ensure that it is followed.

Figure 4.3

4.3 Communication

The communication protocol includes the format, media, and points of control for information disseminated to team members and stakeholders. Key elements of successful communication include consistent delivery, comprehensive horizontal and vertical distribution, and timeliness. The communication protocol also addresses project status meetings, namely, their frequency, duration, location, internal and external attendees, and the standing agenda. One of the first meetings prescribed by the communications plan should be the project kickoff, where the stakeholders and key members of the project team participate in a detailed discussion of the SOW. The communications plan is typically included as *Attachment G* to the SOW.

Figure 4.4a is a sample communications plan. Figure 4.4b depicts typical communication intensity and participants throughout the project phases. Figure 4.5 shows a meeting agenda, and Figure 4.6 depicts meeting minutes. Templates are located in Appendices 22, 23, and 24.

Communications Plan

1. <u>Format for communications</u>. The project manager will distribute guidelines and templates for use in all official project communications, both internal and external.

2. <u>Media</u>. All official project communications will be distributed in written form, either in the text body of an email or as an email attachment. Copies will be retained in the project repository at website www.thisproject.com/documents.   Voice messages will always be supported with emails.

3. <u>Points of control</u>. The project manager or his/her designee will coordinate all communication that involves the entire project team and any external partners. Team leads will ensure that communication within and between teams follows this Communications Plan.

4. <u>Kickoff meeting</u>. The project manager will arrange a project kickoff session involving all stakeholders to review in detail the Statement of Work as soon as it has been completed.

5. <u>Biweekly project team sessions</u>. Every other week, at a time to be determined, the entire project team identified in para. 2 of the Statement of Work will meet to review the following items at a minimum:
   a. overall status of individual action plans
   b. completeness of documentation
   c. risk management
   d. outstanding escalations
   e. status of testing and training, as needed or applicable.

6. <u>Weekly status sessions</u>. Each week, on a day to be determined, the project team leads will meet at least by phone for 30 min. to report on the following:
   a. progress of assigned action items currently due, according to the project plan
   b. new risks
   c. new escalations.

7. <u>Meeting minutes</u>. The project manager (or alternate) will post minutes from all project-level meetings in the web-based repository identified for this project. See SOW Attachment H – Documentation Protocol. Team leads will be responsible for posting minutes to meetings that they conduct.

Figure 4.4a

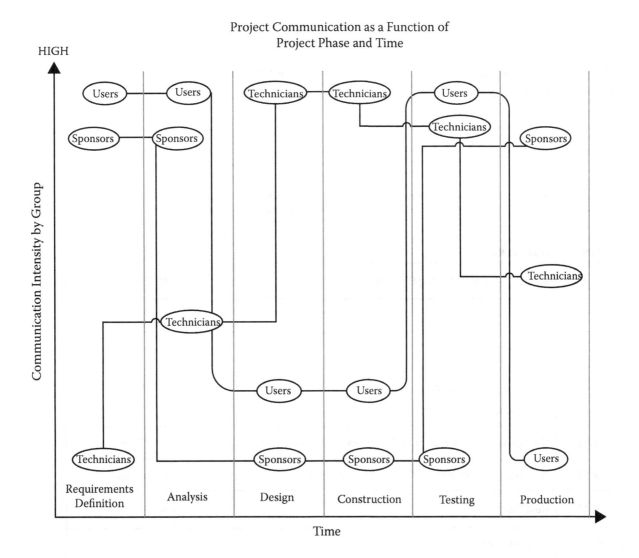

Figure 4.4b

MEETING AGENDA

Project: **National Security Number (NSN) Project**
Purpose: **Kick-Off Meeting**

Date: 7/20/06
Time: 8:00 am – 10:00 am
Place: Conference Room 123

Invitees: Debbie Flora, Ann Garner, Nancy Horla, Andy Jacobs, Gary Rumento, Carol Mason, Lauren O'Brien, Josh Laggy, Karen Randolph, Sharon Stone, Kathleen O'Hara, Mike Zambruski

Time	Agenda Item	Lead
8:00	Introductions and Meeting Purpose	M. Zambruski
8:15	NSN General Overview	L. O'Brien
8:25	NSN Project • Project Charter and Approval • Team Roster and Alternates • Access to Project Documents on Web Repository	L. O'Brien, A. Jacobs
8:45	NSN Business Requirements Document (BRD)—quick review	L. O'Brien
9:15	NSN Statement of Work (SOW)—detailed review	A. Jacobs
9:45	Next Steps	All
10:00	Meeting ends	

NOTES

Figure 4.5

Meeting Minutes

Team Members Attending:	Emmett Walsh, Gary Sensei, Karen Phillips, John Smith, Bob Wang, Mike Zambruski
Team Members Absent:	Stacy Peters
Excused:	George Escobar
Visitors:	Wendy Jersey

Agenda Item	Discussion	Action/Follow-up (assignee & due date)
Process flows ("As Is")	• Services - weekly meetings are still occurring, progress is steady, 4 service categories are completely mapped in an Excel table; next will be flow diagrams • Capital - progress is steady, 40 flows have been mapped, additional 16 pertain to members hospital processes • Supplies - processes mapping is being completed in time for phase 1 implementation (8/14/06)	None
Project teams' Statements of Work	• Services and Capital working teams will modify their S.O.W.s to state that current target dates will be delayed due to redeployment of DBMS expertise to DBMS implementation. However, precise amount of delay is not yet certain, since it depends on when DBMS resources are again available to the Capital and Services working teams. • Key Performance Indicators (KPIs) presented at 8/1/06 Exec. Team Mtg. will be incorporated into S.O.W.s as deliverable metrics to the extent that they apply to the current scope of work. In those cases where applicability of KPI is uncertain for the Program Transformation, or where incorporating them impacts the existing scope of work, guidance from the Exec. Team will be sought.	• Sub-team leaders (John, Karen, Gary, George, Emmett) will modify dates by next meeting (7/13/06).

Figure 4.6

4.4 Documentation

The mode (paper vs. electronic), storage location (physical vs. electronic), and version control of project documents must be formally defined and continuously maintained for easy retrieval and up-to-date accuracy. This pertains to all project definition, planning, and control documents; detailed budget and design records; the issues/risk log; meeting agendas and minutes; training and testing guidelines; workflow diagrams; and any other key reference items. The most efficient way to organize and facilitate this is to establish a project documentation protocol and include it as *Attachment H* to the SOW.

Figure 4.7 shows a standard protocol for storage and retrieval of project documents. A template is included in Appendix 25.

Project Documentation Structure

- **ARCHIVES** — Used to store early or obsolete versions of project files so that we can refer back to them if need be. All of the files in this folder have the prefix "Archive --" appended to the original file name, which makes it easy both to identify them and to keep them distinct from the current version which is posted in one of the other main folders (below).
- **Budget documents** — funds, staff, equipment, supplies
- **Design documents** — conceptual, preliminary, detailed, and final designs
- **Issues and risks Log**
- **Meeting agendas**
- **Meeting minutes**
- **Miscellaneous** — anything not identified elsewhere in this protocol (e. g., procurement plans)
- **Project definition documents**
  - Project Business Case
  - Project Charter
  - Requirements Work Plan (RWP)
  - Business Requirements Document (BRD)
  - Statement of Work (SOW)
- **Project planning and control documents**
  - Project Plans
  - Contingency Plans
  - Escalation Policy
  - Communications Plan
  - Documentation Protocol
  - Project Audits
- **Reference documents** — market studies, quality data, industry information, etc.
- **Training and education documents**
- **Validation and testing documents** — test plans and validation records
- **Workflow documents** — diagrams and charts depicting current and proposed workflows/process flows (both logical and physical)

Figure 4.7

4.5 Testing

Comprehensive validation testing must be planned and conducted against the quality metrics specified in the SOW so that it is absolutely clear when deliverables meet business requirements. Interim verification tests should be developed and conducted at appropriate intervals to gauge progress and mitigate risk. The Testing Protocol is typically included as *Attachment I* to the SOW. Validation testing can take many forms and range from quite basic to very complex in nature. Included in Appendix 18 is a simple template for testing staff readiness prior to implementation of a new business process. Testing that pertains to technical solutions (e.g., medical, telecommunications, information services, etc.) are typically quite sophisticated and under the control of a professional test planning and execution manager.

4.6 Training

Development, delivery, and confirmation of educational material must be assessed at each project phase to determine both the need for and the type of appropriate training. When included it is found at *Attachment J* in the SOW. As is the case with validation testing, training plans can take many forms and range from basic to quite complex. A simple example is presented in Appendix 19 at the end of this book.

Chapter 5

Project Closure

One of the primary reasons for formally ending a project, rather than simply allowing it to disappear from the list of active initiatives, is to help propagate success while hopefully forestalling repeated failure. For this purpose, a formal debriefing session should be included as a milestone at the end of the project plan, and participation should include the entire project team and stakeholders. Figure 5.1 shows an outline for a post-project summary of lessons learned. A template can be found in Appendix 26.

POST-PROJECT LESSONS LEARNED

PROJECT NAME:	PROJECT MGR:	DATE PREPARED:
Project Start Date:	Original Project End Date:	Actual Project End Date:

WHAT CONTRIBUTED TO SUCCESS?
1.

WHAT HINDERED SUCCESS?
1.

PROJECT CHARACTERISTICS

Was the project **planned** properly?	
Were **users** involved in planning?	
Were **risks** identified & managed?	
Were **contingency plans** developed?	
Was the **decision structure** clear?	
Was **communication** timely?	

LESSONS LEARNED

What could have been done differently?	
Why wasn't it done?	
Where will these Lessons Learned be stored for retrieval by others?	

Prepared
by: _____

Date: _____

Figure 5.1

Glossary

- **Business vision** defines the strategic TO BE state of the organization.
- **Business mission** outlines the current AS IS direction of the organization.
- **Business goal** is a major milestone supporting the vision and mission.
- **Business Requirements Document (BRD)** describes in detail *why* a project is needed in order to fulfill a business goal, for example, reduce cost, increase revenue, improve efficiency, optimize customer satisfaction, enhance safety, standardize operations, sharpen compliance, etc. The BRD thoroughly documents these requirements, reflects formal approval by key stakeholders, and thereby specifies the deliverables of a project.
- **Business Requirements Work Plan (RWP)** outlines the effort and resources needed to collect, analyze, synthesize, and validate the business requirements and then formally record them in a BRD.
- **Project charter** is the official authorization for a project in pursuit of the business goal. It documents, at a high level, *what* is to be done.
- **Project Statement of Work (SOW)** is the detailed script for *how* to achieve the project deliverables as specified in the BRD and includes the following information on the project:
  - goal
  - approach
  - tactics
  - priorities
  - progress milestones
  - in scope versus out of scope
  - project team
  - success criteria (deliverables, traceability, quality metrics)
  - assumptions
  - constraints
  - change control process
  - project budget
  - project plan
  - issues/risk management plan
  - issues/risk log
  - escalation policy
  - communications plan
  - documentation protocol
  - test strategy
  - training strategy
  - approvals
- **Stakeholder** is an individual representing any organization that *contributes* to, *benefits* from, or experiences an *impact* from a project (either directly or indirectly). *Contributors* are generally regarded as sponsors. *Beneficiaries* are usually customers and users. *Impacted* parties neither sponsor nor benefit directly from the project; however, they typically need to change one or more of their procedures in order to conform to the improved process created as a result of the project.

Appendix 1

Hierarchy of Enterprise Targets

PROJECT MANAGEMENT IN THE OVERALL ENTERPRISE ENVIRONMENT

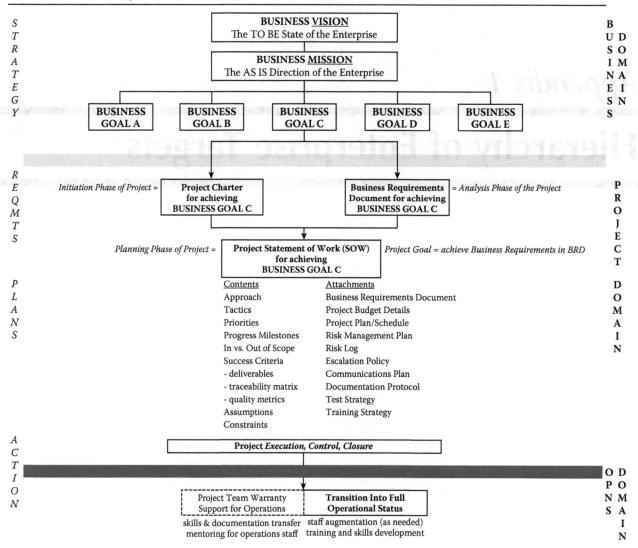

Appendix 2

Project Assessment Form

Date: mm-dd-yyyy	Sponsor/Requestor: <<name>>	Project Name: <<name>>	Prepared by: <<name>>
Project Details	*Client Expectations*	*Business Analysis & Project Management Needs*
WHAT? ■ objectives ■ deliverables ■ assumptions ■ constraints		
WHY? ■ strategic goal(s) ■ business requirements ■ acceptance criteria		
WHEN? ■ key milestones ■ phases ■ status		
WHO? ■ stakeholders (orgn & indiv) ■ PMs ■ core team ■ related projects		
HOW? ■ financials ■ resources ■ issues/risks ■ communication ■ change ctrl ■ escalation ■ testing ■ training ■ lessons learned ■ documentation		

NOTES:

Appendix 3

Project Initiation Document Workflow

Project Initiation Documents

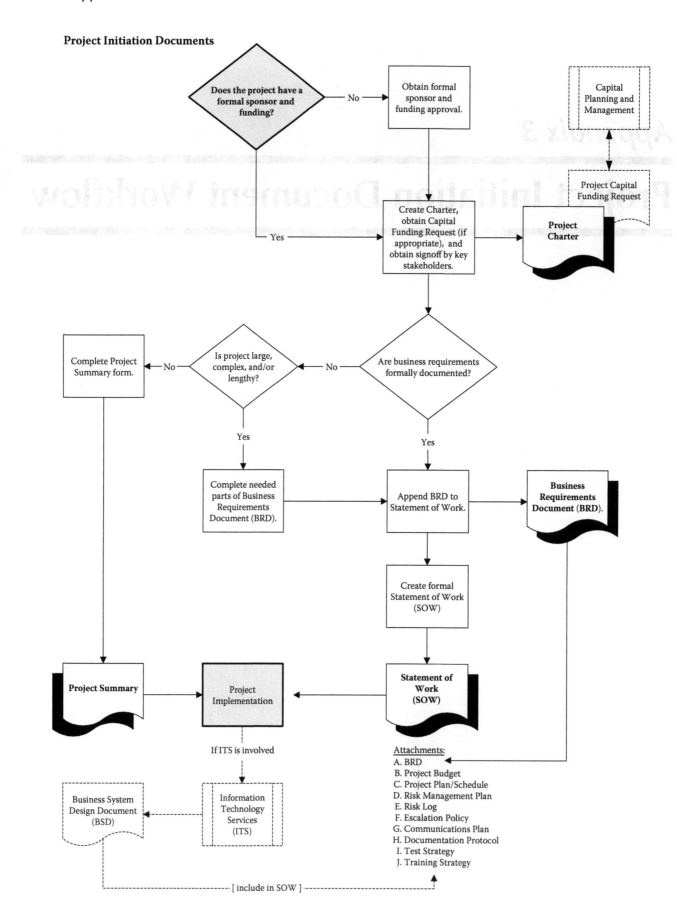

Appendix 4

Project Charter Template

Date:	Initiated by:	
Project Summary Details		
Project Name:		
Project ID:		
Project Priority:		
Customer Name:		
Project Start Date:		
Planned Project End Date:		
Approved Budget:	*<<initial estimates are acceptable at this stage, to be refined as business analysis is completed>>*	
Project Staffing Level (Total Person Months):	*<<initial estimates are acceptable at this stage, to be refined as business analysis is completed>>*	
Project Personnel		
Project Sponsor(s):		
Business Owner(s):		
Project Manager:		
Other Key Personnel:	*<<typically included here are managers or key subject matter experts who will be instrumental to the success of the project>>*	
Scope and Objectives:		
<< The project scope and objectives are presented here at a high (executive) level. If a separate document contains that information, it can be embedded here.>>		
Organizational Relationships (Roles and Responsibilities):		
<<All organizations that will either contribute to this project or be impacted by it should be listed, together with the title of the person who will represent that organization for this project. This ensures that the organizational breadth of this project is clarified at the very beginning.>>		
Key Dates or Milestones:		
<<Very high level milestone dates are listed, to establish the general timeframe for the project.>>		
Approvals:		
Date:	Signature:	*<<electronic approval is acceptable if a dated e-mail is referenced>>*
Date:	Signature:	

Appendix 5

Completed Project Charter

Date: June 2006	Initiated by: John Smith, Vice-President
Project Summary Details	
Project Name:	**National Security Number (NSN) Project**
Project ID:	06-06-013
Project Priority:	HIGH
Customer Name:	Eastern Region Business Unit
Project Start Date:	6/2006
Planned Project End Date:	6/2007
Approved Budget:	$1,310,500
Project Staffing Level (Total Person Months):	18.5
Project Personnel	
Project Sponsor(s):	John Smith
Business Owner(s):	Terry O'Rourke
Project Manager:	Harry Easter
Other Key Personnel:	Mike Zambruski, IT, Finance. Refer to NSN Project Team roster.

Scope and Objectives

The NSN Project Team is responsible for addressing and implementing an upgraded customer service that ensures direct person-to-person involvement when the company is contacted for any post-purchase reasons.

Specific objectives include:

- Review existing policies, procedures, and transactions.
- Identify and analyze gaps between requirements and current practices.
- Develop unit-wide plans to achieve the desired benefits specified by the corporate policy.
- Ensure vendor partners are involved in the entire planning and implementation process.
- Document all analysis and rationale for final solution(s).
- Determine key steps to implement the service in the Eastern Region Business Unit.
- Assign a full-time project manager to coordinate all constituent projects and related activities.

Organizational Relationships (Roles and Responsibilities)

The vice-president will have lead responsibility and accountability for this project. This determination has been made because the project is a major competitive issue.

Key Dates or Milestones

7/2006—Kickoff meeting with all stakeholders
1/2007—First customer tests

6/2007—Final implementation throughout region

Approvals

Date: 6/16/06   Signature: John Smith, Vice-President, Eastern Region Business Unit (*email attached*)
Date: 6/16/06   Signature: Patrice Longley, Executive Vice-President (*email attached*)

Appendix 6

Project Summary Template

Project Title:	Project Summary Author:
Project Control No.:	Date Prepared:
Client:	Revised on:
Sponsor:	Revision No.:

1. WHAT?

a) Project Objectives—including priorities (attach Project Charter)
b) Project Deliverables—both in scope and out of scope
c) Assumptions
d) Constraints

2. WHY?

a) Strategic goals supported by project
b) Business requirements to be satisfied by project deliverables

c) Acceptance criteria (including quality standards and metrics)

3. WHEN?

a) Key milestones (attach detailed project plan/schedule)

b) Phases (if applicable)

c) Status—i.e., is it already under way?

4. WHO?

a) Project management

b) Stakeholders (including organization and title)

c) Project team (include or attach team roster with roles, responsibilities, contact information, and alternates)

d) Related projects

5. HOW?

a) Financial analysis (including project budget)

b) Key resource requirements (including labor, equipment, supplies, information)

c) Issues and risks (attach issues/risk log)
d) Communication plan
e) Change control process
f) Escalation policy
g) Testing, validation, and training required
h) Application of former lessons learned
i) Documentation standards

APPROVALS

NAME	TITLE	APPROVAL	DATE
		/s/ or attach e-mail	
		/s/ or attach e-mail	
		/s/ or attach e-mail	
		/s/ or attach e-mail	

Appendix 7

Business Requirements Work Plan Template

Business Requirements Work Plan (RWP)

<< Project Name >>

Date prepared (or updated):	
Prepared by:	
Project sponsor:	
Project customer:	
Project manager:	
Business analyst:	
Project ID:	

1. Requirements Work Plan (RWP) Objectives

The objective of this document is to provide details on how the analysis phase of the overall project will be conducted. It describes the methodology, plans, and resources to be employed in creating a Business Requirements Document (BRD), which will ultimately become the blueprint for all subsequent project work. When completed, this RWP should actually be integrated with the project plan to reflect the work to be done during the analysis phase of the project.

2. Deliverables

The primary deliverable of the effort described in this RWP is production of a formal Business Requirements Document, which will then serve as the key project reference for developing and implementing appropriate solutions.

3. Scope

The scope of this document is limited to efforts directed at producing a BRD for the project as it is defined by its charter. It is *not* the scope of this RWP to actually provide conclusions or recommended solutions. This is a planning document intended to clarify the level of effort, funds, and time needed to conduct the analysis required to position the project for success.

4. Participating Stakeholders

Specify which representatives from each of the following areas will be contacted as deemed necessary to provide a comprehensive understanding of the business requirements pertaining to this project:

- Sponsors, who contribute directly to the project in terms of resources (funds, labor, equipment, supplies, information, etc.). **<< provide names and/or organizations >>**
- Beneficiaries, who benefit directly from the project's results. This typically includes customers and users. **<< provide names and/or organizations >>**
- Affected parties, who neither sponsor nor benefit directly from the project but who will be impacted by its outcome. Typically this includes individuals or organizations who will need to make some change to their processes or tasks as a result of the project, but who may not see an immediate benefit. (An example would be upgrades to computer workstations throughout an entire organization.) **<< provide names and/or organizations >>**

5. Resource Requirements

Specify the staff, equipment, and external support (as applicable) that will be needed to perform the analysis leading to the BRD. Keep in mind that this is purely the analysis team, not the overall project team.

6. Assumptions

Enumerate anything that might constitute a risk and therefore warrant close oversight and continual updates during the analysis phase, for example, project priority, executive support, sufficient funding, available staffing, and time.

7. Dependencies

Describe the relationship that this project has to other projects or an overall program, and how these dependencies may affect the timely, accurate completion of the BRD.

8. Constraints

List any known limitations on conducting a proper business analysis for the project, for example, schedule, budget, labor resources, partner availability, etc.

9. Attachments

List and attach appropriate supporting and background documents, for example:

- The project charter
- The RWP analysis team roster
- The RWP budget
- A work breakdown structure (WBS) showing RWP activities
- Any other items that clarify the efforts to be undertaken to produce the BRD

10. Approvals

Business Analyst	Date
Project Manager	Date

Appendix 8

Business Requirements Document Template

Business Requirements Document (BRD)

<< Project Name >>

Date prepared (or updated):	
Prepared by:	
Project Sponsor:	
Project Customer:	
Project Manager:	
Business Analyst(s):	
Project ID:	

Table of Contents

BRD Revision Log

Revisions to the Business Requirements Document should reflect at least the following information:

Date Change Proposed	Description of Change	Reason for Change	Date Change Approved

1. Project Goal, Objectives, Scope

Describe those strategic goals and objectives that the project under consideration is intended to fulfill. If a strategic mission and vision document or statement is available, it should be included here for reference. Also describe what is clearly out of scope for this project and, therefore, for this BRD.

Keep in mind the audience for whom this is written, to ensure that the proper level of detail is maintained throughout. An organizational chart for the affected business unit(s) should be included.

2. Background

Enter sufficient detail to orient the reader as to the context surrounding this initiative—especially whether this is a phase of a multiphase project, whether prior efforts have failed (and why), and whether subsequent activity or phases are planned.

3. Business Analysis Process

Describe the process(es) employed for producing this BRD. Specifically, indicate whether research, interviews, surveys, focus groups, or the like were employed. Also indicate whether business use cases were employed and where they are available for review.

4. Stakeholders

Indicate the names of individuals who were contacted for this BRD, classify them in the following three stakeholder categories, and specify which have approval authority for this BRD:

- Sponsors, who contribute directly to the project in terms of resources (funds, labor, equipment, supplies, information, etc.).
- Beneficiaries, who benefit directly from the project's results. This typically includes customers and users.
- Affected parties, who neither sponsor nor benefit directly from the project, but who will be affected by its outcome. Typically this includes individuals or organizations who will need to make some change to their processes or tasks as a result of the project, but who may not see an immediate benefit. (An example would be upgrades to all computer workstations throughout an organization, which includes those recipients who do not necessarily need the upgrade.)

5. Business-Level Requirements

Identify the *business-level performance improvements* that are needed and that the organization expects the project to satisfy. Examples would be reduced cost, increased customer satisfaction, enhanced efficiency, better quality, stricter regulatory compliance, increased marketplace performance, etc. Include all long-term (strategic), short-term (tactical), and on-going (operational) factors that help to clearly define these business-level requirements.

6. User Requirements

Describe the *performance improvements to individual business processes and systems* that will help users more successfully perform their tasks as a result of the project. Examples would be "easy access to customer support any time of day" or "a simpler way for customers to exchange information with the organization". These requirements are typically described from the user's point of view and in nontechnical user terms. As a result, they are often assembled in a narrative form and supported by various graphic models, such as work flow charts, use case diagrams, activity (swim lane) diagrams, and the like.

7. Functional Requirements

Specify the *operational behavior of business processes and systems* that will lead to satisfying the user requirements described above. In particular, this includes how users will interact with manual or automated systems that will be part of the project deliverables. The following are examples: providing clients with 24-hour Internet access to customer service staff (instead of having to make telephone calls only during the day); providing patients with CDs that contain their x-rays (instead of large films that require special light readers); and enabling students to complete only one application form for any state college in the same state (instead of a unique form at each institution). All three instances describe functional interaction with the organization's business processes that will satisfy the above user requirement for simpler transactions and in turn meet the above business-level requirement of improved customer satisfaction.

Detailed technical specifications are *not* included here, because they will be developed later in the project during the solution design and construction phases. However, this section should contain sufficient information on the needed functionality so that solutions being developed and tested during the project can be repeatedly compared against the users' expectations.

8. Nonfunctional Requirements

Nonfunctional requirements include anything that enhances the success and value of the project deliverables to the users. This includes training, documentation storage and retrieval, security, service-level agreements, integration with other business processes, and any other aspects of successful performance that users typically take for granted and do not request directly.

9. Reporting

Specify the content, format, source, frequency, and intended audience of any and all reports that are expected as deliverables of this project. These can include regularly scheduled reports, ad hoc inquiries, and audit reports.

10. Assumptions

Beginning with the most basic assumptions—such as sufficient funding, staffing, and executive support—this section specifies all vital areas of the project that are presumed to be constant and therefore warrant close oversight and continual updates.

11. Dependencies

Specify the relationship that this project has to other projects, to phases of an overall program, and to any upstream or downstream elements that the project can affect or be affected by.

12. Constraints

Any limitations on the project, for example, schedule, budget, labor resources, partner availability, etc., must be listed here. They set the boundaries within which the project is expected to operate.

13. Risk Identification, Management, and Escalation

Clearly delineate the process for identifying, managing, escalating, and retiring risks that can impact the project. Keep in mind that risk equals uncertainty, and that all uncertainty, whether small or large, deserves identification and management as the project is formulated and then executed.

14. Technology Considerations

Existing technology and its impact on the envisioned project should be outlined here, together with any newer technology to be considered.

15. Solution Considerations

Although the BRD is not intended to provide final solutions—as that is the domain of the project itself and its Statement of Work—nevertheless, any information available on pilot programs, proof of concept work, or rejected solutions should be included here. This will ensure that such historical knowledge is part of future project work, thus forestalling possible duplication of effort.

16. Change Management

Once the final BRD is approved with signatures from key stakeholders, changes to it should only occur under strict change management control. Details of such a control process should be reflected here, including what constitutes a change, as well as how its impact should be evaluated prior to seeking approval from the original approvers of the BRD.

17. Attachments

For completeness and convenience to readers, include in this section all appropriate supporting documents, such as the Project Charter, Business Case, Requirements Work Plan, Process Modeling Diagrams, etc.

18. Approvals

1. Prepared by	
Business Analyst	Date
2. Approved by	
Project Manager	Date
3. Approved by	
Customer	Date
4. Approved by	
Sponsor	Date

19. Glossary of Key Terms in BRD

Appendix 9

Sample of a Completed Business Requirements Document

National Security Number (NSN) Customer Service Project

Date prepared:	7/1/2006
Prepared by:	Brian Terry
Project Sponsor:	John Smith
Project Customer:	Terry O'Rourke
Project Manager:	Harry Easter
Business Analyst:	Brian Terry
Project ID:	06-06-013
Last Update:	9/6/2006

Table of Contents

BRD Revision Log

Date Change Proposed	Description of Change	Reason for Change	Date Change Approved
9/1/06 (rev 1.0)	Harry Easter (Project Mgr)	addition of nonfunctional requirements in section 8	9/6/06

1. Project Goal, Objectives, Scope

The NSN Project Team is responsible for addressing and implementing an upgraded customer service that ensures direct person-to-person involvement when the company is contacted for any post-purchase reasons.

Specific objectives include:

■ Review existing policies, procedures, and transactions
■ Identify and analyze gaps between requirements and current practices
■ Develop unit-wide plans to achieve the desired benefits specified by the corporate policy
■ Ensure vendor partners are involved in the entire planning and implementation process
■ Document all analysis and rationales for final solution(s)
■ Understand and determine key steps to implement NPN customer service in the Eastern Region Business Unit

The intended audience for this BRD is all business and technical staff who either contribute to, benefit from, or are affected by the creation and implementation of this new higher-grade customer service.

2. Background

Research has demonstrated that the explosive growth of the Internet and the concomitant rise in security risk have led to a proliferation of security firms that promise protection that often is not delivered. In response to these factors, we created the National Security Number system to register, screen, and validate the credentials of these firms. This project furthers those efforts by upgrading the support provided to customers once they purchase these services.

We have determined that customers expect personalized care when contacting us for inquiries, complaints, and status updates on typically sensitive issues. Therefore, contrary to most modern post-sale services that use prerecorded responses activated by telephone number pads, our service will stress the added value of immediate human interaction with our clients.

3. Business Analysis Process

The business needs and user needs were verified through extensive but carefully targeted interviews and focus groups, to ensure comprehensive input into the analytical process. Information thus gathered was further categorized according to functional, nonfunctional, and system requirements, which will then become the blueprint for system specifications to support development of final solutions. All documented findings and analysis will be kept in a Web-based repository for version management and efficient retrieval.

4. Stakeholders

Final approval of the Business Requirements Document resides with John Smith, VP, who is the project sponsor. Primary users are identified as customer service representatives (CSRs) who handle calls coming from clients. Secondary users include the company financial, contracting, marketing, and legal staffs, each of whom in some way supports the CSRs in successfully responding to customer inquiries.

Other key stakeholders include:

- Farley Cooke, chief technology officer
- Gary Sorrenson, senior VP, compliance
- Bruce Petersen, security director
- Patrick Connelly, contacting office director
- Michael Zambruski, director of project management

5. Business-Level Requirements

The following business requirements must be satisfied by this project:

- Regulatory—Develop processes that ensure the highest possible compliance with privacy regulations that apply to the sensitive data typically collected from individual customers.
- Strategic—Increase our NSN service quality as the de facto industry standard for overseeing Internet security firms, and in this way further solidify our dominant market share of this service (currently at 60 percent).
- Tactical—Create a responsive and professional customer service cadre that epitomizes the industry-leading position that we hold for this service.
- Operational—Ensure that recruitment, training, and quality control practices are in place to guarantee the highest level of personal service to customers in an on-going manner.

6. User Requirements

Customer service representatives need access to all corporate systems supporting this service, and such access must be easy for staff to use when in stressful communication with an agitated client.

7. Functional Requirements

Functional requirements supporting the above user requirements include quick data retrieval and correlation, interactive plain-language query procedures, a full history of prior inquiries from the subject customer, and supervisory access within 30 seconds for incidents requiring escalation for resolution.

8. Nonfunctional Requirements

a. Although many legacy and state-of-the-art corporate systems will be employed in providing this service, its success relies on full integration across all such systems. This will ensure optimal accuracy of customer data presented to the CSRs when they respond to and resolve inquiries in real time.

b. Past experience with sensitive customer-service activities strongly supports a Web-based graphical user interface (GUI) as the preferred method for CSRs to interact with corporate systems needed for professional and timely responses to customers.

   c. Typical user ID and password access will be required for authorized on-site CSRs to use the system(s) supporting this service. Remote access will be provided on a controlled as-needed basis, supported by Virtual Private Network (VPN) technology as is done for similar corporate resources.

   d. All data systems supporting this newly upgraded service must be available from 0700 to 2000 Eastern Time on workdays and 0900 to 1600 on Saturdays, Sundays, and holidays. Also, it is anticipated that as many as six full-time CSRs may concurrently access the needed systems.

   e. CSRs will need the ability to input key customer data (name, account number, address, phone numbers, etc.), as well as a free-form text entry describing the nature of the inquiry—all done through the GUI portal suggested above.

   f. System downtime or outages must not exceed 15 minutes during prime operating time, as defined above.

   g. All customer data supporting this service will be backed up nightly, with archives produced monthly and stored at an off-site service provider for 3 years.

   h. This newly designed customer service is intended for indefinite application, with improvement evaluations to be scheduled and conducted on an annual basis.

   i. All CSR processes, scripts, system access training, and customer interaction skills will be documented prior to implementation. Further, the human resources department will participate in ensuring that such documentation supports the training and certifications described below.

   j. The basic purpose of this improved customer service is to enhance the customer experience to the point where our firm is clearly recognized as the industry leader. To that end, CSRs participating in this effort will receive extensive system and customer-interaction training, supported by the human resources and marketing departments. They will also be certified as having completed such training before they begin participation in the service.

   k. Usability of the GUI or equivalent described above will be extensively tested prior to operational implementation. Such usability testing will ensure uniform application and success by all CSRs, whether neophytes or veterans.

   l. All screen text will maintain the same look and feel as that which is found on other corporate Web sites, both internal and external.

9. Reporting

Results of customer service transactions must be summarized for timely access by senior management, who will use such reported data to gauge progress with the planned improvements in support of the customer. Detailed contents, frequency, and format of such reports are appended at the end of this BRD and include prototype reports for illustration.

10. Assumptions

   a. All involved project team members are present at the kick-off meeting and throughout the project.

   b. Key stakeholders and sponsors provide timely feedback for issues and status.

   c. Necessary funding is available for extra resources if required.

11. Dependencies

Key dependencies are shown in the detailed project plan.

12. Constraints

   a. This support service must be ready for implementation within 4 to 6 months of project initiation, to coincide with the launch of the new NSN service.

   b. We must ensure that filling key CSR positions for this service does not cannibalize staff from our other primary customer-facing services.

13. Risks Identification, Management, and Escalation

Project risks will be identified and managed in accordance with the following template:

Risk	Likelihood	Impact	Owner	Contingency
Lack of history with new service will put CSRs under constant stress to deal with new problems and concerns from customers.	Medium probability	High	Marketing [George Still] & Human Resources [Alexandra Hollcomp]	Marketing & HR must work together in designing scripts and FAQs for CSRs to follow during training and actual service implementation.
Limited compatibility among legacy and current state information systems will result in diminished responsiveness from CSRs dealing with customers in real-time situations.	High probability	High	Information Technology[Geoff Hartman]	1. Develop comprehensive map of interacting systems. 2. Prioritize above systems from most to least critical for success. 3. Outline strategy for providing integrated access to high-priority systems by CSRs.

14. Technology Considerations

The new customer service must provide clients with comprehensive and immediate support on all inquiries, regardless of how they are transmitted to the company (phone, e-mail, fax, or surface mail). CSRs will therefore need immediate and efficient access to all pertinent customer account information on a real-time basis. This will require integration of many databases that up to now have functioned independently. It may also require development of a front-end access method for CSRs, similar to the functionality provided by a Web-based GUI.

15. Solution Considerations

Because this service closely resembles similar customer-facing services provided by the company, it is expected that those current successful services will serve as the model for this new initiative.

16. Change Management

Substantive changes in scope, schedule, funding, quality, or risk must undergo scrutiny and approval by the key stakeholders before being implemented. The standard **Project Change Request Form** will be used for this purpose.

17. Attachments

    A.   IT Systems Diagrams

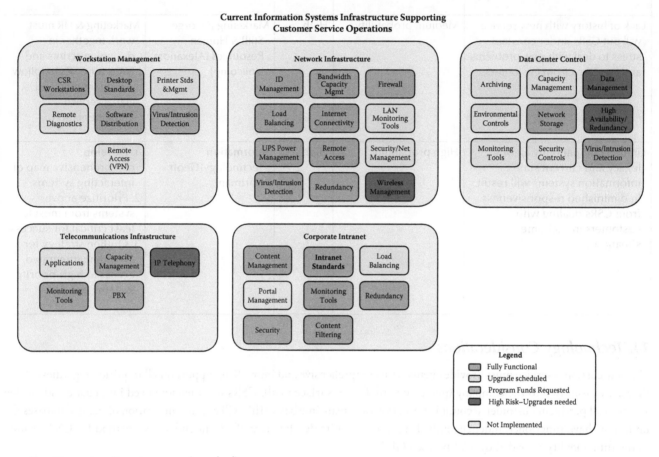

    B.   Reporting Requirements (attached)
    C.   Project Plan (attached)
    D.   Project Change Request form (attached)

18. Approvals

This document has been approved as the official Business Requirements Document for the National Security Number (NSN) Customer Service project and accurately reflects the current understanding of business requirements. Following approval of this document, requirements changes will be governed by the project's change management process, including impact analysis as well as appropriate reviews and approvals, under the general control of the Project Plan and according to company policy.

Name	Title	Function	Date
Brian Terry	Business Analyst	Author of BRD	July 8, 2006
Terry O'Rourke	Director, Customer Service	Business Owner	July 10, 2006
John Smith	Vice President	Sponsor	July 11, 2006
Gary Sorrenson	Sr. VP, Compliance	Stakeholder	July 13, 2006

Farley Cooke	Chief Tech Officer	Stakeholder	July 13, 2006
Michelle Ritz	VP, Human Resources	Stakeholder	July 20, 2006
Josef Stakowski	VP, Marketing & Sales	Stakeholder	July 21, 2006
Michael Zambruski	Director, Proj Mgmt Off	Advisor	July 10, 2006
Harry Easter	Project Manager	Project Mgmt	July 9, 2006

Please note: Electronic signatures are acceptable if verified by documents such as e-mails that clearly indicate the approving person and date. Such e-mails are filed with the project documentation in a Web-accessible repository.

19. Glossary of Key Terms in BRD

Customer Service Representative (CSR)—one of the trained staff dedicated to answering customer inquiries, whether such inquiries are by phone, mail, e-mail, fax, or personal appointment.

Graphical User Interface (GUI)—an Internet Web-based display that provides users with full-screen capabilities to enter and access data as well as link to other pages and sites, all with a minimum of keystroke entry.

National Security Number (NSN)—a standard unique identifier for security providers. The NSN is a seven-digit, intelligence-free numeric identifier. Intelligence free means that the numbers do not carry information about the providers, such as the state in which they practice or their specialization.

Standard Transaction—includes background checks, nationality inquiries and responses, and criminal history inquiries and responses.

Appendix 10

Statement of Work Template

Statement of Work

<Project Name>	
<Project Duration>	
<Project Manager>	
<Contact Information>	
Date Prepared (or Updated):	
Prepared by:	
Document File Name:	
Document FileLocation:	
Project ID:	
Project Sponsor:	
Project Customer:	
Business Analyst(s):	

Version History			
Date Change Proposed	*Description of Change*	*Reason for Change*	*Date Change Approved*

1. Project Description

1. **Goal**—summary of how the project will achieve the formally documented business requirements, to include a description of the following:

   a. **Business strategic goal(s) and objective(s) supported by this project**—if there are none, then the legitimacy of the project should be questioned and explained.

   b. **Significant risks**—Any enterprise-level factors (such as competition, regulations, economics, etc.) jeopardizing project success.

2. **Approach**—high-level explanation of how the project will be organized and managed, with specific reference to the processes that will be followed, such as:

   a. **Normal application of the Standard for Enterprise Project Management**, with careful, deliberate completion of the supporting documents in the Attachments section of this Statement of Work.

   b. **Crisis application of the Standard for Enterprise Project Management**, with a justification for the time sensitivity, its impact on team members' priorities, and the effect of accelerated progress on quality.

   c. **Application of prior lessons learned**, including the approaches, techniques, or improvements to be applied to this project based on experience in past projects.

3. **Tactics**—a list of key steps needed to initiate and control the project, for example, formulate project team, prepare preliminary plans, conduct kick-off, convene regular status meetings, etc. If this is a *crisis*, details of how it will be managed should appear here.

4. **Priorities**—a high-level list of deliverables for this project, shown in order of priority (with details provided in the comprehensive project plan/schedule at Attachment C).

5. **Milestones**—specific points in time targeted for measuring project progress.

6. **Out of scope**—clear statement of what is *not* included in this project.

2. Project Team

Name	Title	Department (or Firm)	Area of Specialty	Role/ Responsibility	Phone Contact	Alternate Representative	Alternate's Phone

3. Success Criteria

a. **Key deliverables**—details of tangible, measurable results that are to be achieved as a result of this project.

b. **Traceability to Business Requirements Document (BRD)**—correlation between each deliverable and the corresponding need stated in the BRD. An example is shown below:

Business Requirement	User Requirement	Project Deliverable
B-1. Improve competitive position by providing unusually superior service to customers after they complete a purchase.	U-1. Customer service reps (CSRs) must provide direct person-to-person support when customers contact the company for any post-purchase reasons.	S-1. Implement the new customer service processes and support systems by the summer of 2007 with a budget of $1.31 million.

c. **Quality metrics**—measurable and testable characteristics that serve as customer acceptance criteria for each deliverable.

4. Assumptions

The source and reliability of each assumption should be included here.

5. Constraints

This includes the source and contact information for each constraint (budget, scope, schedule, quality).

6. Change Control Process

The formal, documented process that includes all stakeholders whenever formal or informal changes to scope, budget, schedule, or acceptance criteria are proposed. The Standard for Enterprise Project Management contains a **Project Change Request Form** template to facilitate this process (Appendix 20).

Attachments

A. **Business Requirements Document**
B. **Project Budget Details**—to include R.O.I., payback period, cost-benefit analysis, expense reduction, revenue improvement, make-buy analysis, opportunity cost, etc., as appropriate.
C. **Project Plan/Schedule**
D. **Risk Management Plan**
E. **Risk Log**
F. **Escalation Policy**
G. **Communications Plan**
H. **Documentation Protocol**
I. **Test Strategy**
J. **Training Strategy**

Approvals

Authorized Approver Name:	Approver's Department	Date Approved:
	Executive Sponsor	\<electrronic record – e.g., email– is acceptable\>
	Supporting Organizations(s)	\<electrronic record – e.g., email– is acceptable\>
	Project Manager	\<electrronic record – e.g., email– is acceptable\>
	Other . . .	\<electrronic record – e.g., email– is acceptable\>
	Other . . .	\<electrronic record – e.g., email– is acceptable\>

4. Assumptions

The source and reliability of each assumption should be included here.

5. Constraints

This includes the source and contact information for each constraint (budget, scope, schedule, quality).

6. Change Control Process

The formal, documented process that includes all stakeholders whenever formal or informal changes to scope, budget, schedule, or acceptance criteria are proposed. The Standard for Enterprise Project Management contains a **Project Change Request Form** template to facilitate this process (Appendix 20).

Attachments

A. Business Requirements Document
B. Project Budget Details—to include ROI, payback period, cost-benefit analysis, expense reduction, revenue improvement, make-buy analysis, opportunity cost etc. as appropriate.
C. Project Plan/Schedule
D. Risk Management Plan
E. Risk Log
F. Escalation Policy
G. Communications Plan
H. Documentation Protocol
I. Test Strategy
J. Training Strategy

Approvals

Authorized Approver Name	Authorizes Department	Date Approved
	Executive Sponsor	electronic record - appropriate if acceptable
	Supporting Organization	electronic record - e.g. email; receipt date
	Project Manager	electronic record - e.g. email - is acceptable
	Other	electronic record - e.g. email - is acceptable
	Other	electronic record - e.g. email - is acceptable

Appendix 11

Sample Completed Statement of Work

Statement of Work

Emergency Communications System Project	
September, 2006–February, 2008	
Project Manager	**Michael S. Zambruski 203-555-1212**
Date prepared (or updated):	**September 12, 2006**
Prepared by:	**Michael S. Zambruski**
Document file name:	**SOW v3.doc**
Document file location:	**shared storage drive TOWNPUBSAFETY01**
Project ID:	**06-331**
Project Sponsor:	**James Kirkpatrick (Mayor)**
Project Customer:	**Municipal Emergency Preparedness Office, Local Law Enforcement Jurisdictions, five local medical facilities**
Business Analyst(s):	**Brian Lawson**

Change Control			
Date Change Proposed	Description of Change	Reason for Change	Date Change Approved
10/5/06	funding level increased	funds approved for rehearsals & drills	10/11/06
10/4/06	target milestone dates	new dates proposed and accepted	10/12/06

1. Project Description

a. Goal—Organize, coordinate, and control the relocation and resumption of the following services from their current base in the town courthouse to the new facilities provided by the Municipal Expansion capital project:

- Emergency public safety communications, including radio, wireless telephone, and digital data transfer
- Medical emergency transportation to four local hospitals and one regional center, including both ground and air routes

(1) This project will support the strategic goal of providing state-of-the-art public safety and medical response services to the community.

(2) A significant risk stressed by the senior management team involves the potential disruption to continuous service that could arise during the transition from the old to the new location.

b. Approach—A distributed management approach will apply, as follows:

(1) Central coordination will be provided by Michael Zambruski (project management).

(2) Senior executives from the Municipal Emergency Preparedness Office, local law enforcement jurisdictions, and affected medical institutions will coordinate the following activities within their respective organizations:

- Comprehensive *transition action plans* for operations during the pre-move, move, and post-move periods. A template is included in Attachment C. These must be in line with the overall logistics and construction schedules.
- Coordination with *interdisciplinary and support services* for both the pre-move and post-move periods. See the project team roster and Task-Team Matrix in section 2 Project Leadership Team in this SOW.
- *Workflow diagrams* (for both existing "AS IS" workflows in the current space, as well as planned "TO BE" workflows in the new space).
- Identification and management of ongoing *issues and risks* (see Attachment E).
- *Contingency plans* for worst-case scenarios.

c. Tactics

(1) A **Transition Leadership Team** will consist of the senior executives referenced in section 2 Project Leadership Team in this SOW.

(2) An **Operations Team** will be formed for each area being moved and will consist of key management and staff personnel who will be responsible for coordinating the details of the move.

(3) **Status meetings** will be held biweekly to ensure steady, efficient progress. More frequent sessions will occur as deemed appropriate.

(4) Each area affected by the move will prepare a comprehensive transition **action plan** to be used as the primary control and tracking device throughout the project. See Attachment C.

(5) A formal **escalation** process is included in Attachment F.

(6) Project **documentation** will be stored on a shared drive with access limited to team members and their delegates.

(7) A formal **communication** plan has been developed and is being managed by the Emergency Preparedness Communications Department.

(8) **Risks** and issues will be formally identified and managed by each area being moved.

(9) **Testing** will occur via rehearsals using mock facilities.

(10) **Training** on new equipment will be coordinated by Municipal Engineering.

d. Priorities are reflected in the sequence for moving equipment, furnishings, and functionality, as follows (dates are preliminary, and workflows plus contingencies are critical):

(1) Wednesday, 1/23/08—Emergency Preparedness Administration moves into new space.

(2) Saturday, 1/26/08 (hourly schedules to be developed)

- Emergency communications equipment will relocate from the old into the new space.
- Ambulance service relocation to the new space will be complete.
- Medical clinic will reopen in new space.

(3) Sunday, 1/27/08 (hourly schedules to be developed)

- All new emergency calls will coordinate through the new location, to include local hospital communications (ground and air).
- New helipad will open.

e. Milestones—See project plan in Attachment C.

f. Out of scope—The following activities are outside the scope of this project but will require coordination to ensure overall success:

(1) Emergency vehicle issues and tasks not related to communications or transport routes.

(2) Consolidation of ambulance services into new space.

(3) Completion of medical clinic in new space.

(4) Readiness of new helipad.

2. Project Leadership Team

Department	Representative	Contact	Alternate	Contact
Municipal Emergency Communications	Cicely Tangery	123-456-7890	Ed Harper	123-456-7891
Police Precinct No. 1	Sgt. Edward Jenkins	123-456-7899	Off. Tom Skills	123-456-7898
Police Precinct No. 2	Lt. Emily Harris	123-456-9876	Sgt. Mike Riley	123-456-9871
City Memorial Hospital	Grayson Edwards	123-654-9971	Dr. Leonard Pitt	123-654-9833
County General Hospital	Paul Jenkins	122-344-6677	Sandy Major	122-344-6678
University Medical Center	Dr. Lauren Fields	123-666-7777	Dr. James Kirch	123-666-7801
Westside Medical Center	Katherine Thompson	122-444-5555	Tiffany Elgin	122-444-5556
St. Mary's Hospital	Sister Mary Pancratia	122-555-6666	Bud Blinds	122-555-6699

EMERGENCY COMMUNICATIONS SYSTEM RELOCATION
Task-Team Matrix

Project Sponsor = James Kirkpatrick (Mayor)

Project Manager = Michael S. Zambruski

Michael S. Zambruski
123-456-7890

Each cell that intersects with the three affected organizations (columns) and the various interdisciplinary & support services (rows) should show a date when coordination was done, plus individuals involved. If not applicable, show N/A.

INTERDISCIPLINARY & SUPPORT SERVICES	MUNICIPAL EMERGENCY PREPAREDNESS OFFICE				POLICE PRECINCTS # 1 & 2				4 LOCAL & 1 REGIONAL HOSPITALS				City Memorial / County General / University Med Ctr / Westside Med Ctr / St. Mary's
	Cicely Tangery	Ed Harper	Cicely Tangery	Ed Harper	Sgt. Jenkins Lt. Harris	Off. Skills Sgt. Riley	Off. Skills Sgt. Riley	Sgt. Jenkins Lt. Harris	G. Edwards P. Jenkins Dr. Fields K. Thompson Sister Pancratia	Dr. Pitt S. Major Dr. Kirch T. Elgin B. Blinds	Dr. Pitt S. Major Dr. Kirch T. Elgin B. Blinds	G. Edwards P. Jenkins Dr. Fields K. Thompson Sister Pancratia	
	Project Plan	Workflow Diagrams	Issues & Risks Log	Contingency Plans	Project Plan	Workflow Diagrams	Issues & Risks Log	Contingency Plans	Project Plan	Workflow Diagrams	Issues & Risks Log	Contingency Plans	
Blood bank													
Central Dispatch/Transport													
Communications													
Disaster planning													
Facilities/maint. & engrg.													
Fire Safety													
Human Resources													
Information Services													
Interpreters													
Life Flight													
Material Management (supplies)													
Patient Access Services													
Patient Safety													
Patient Liaison													
Pharmacy													
Psychiatry													
Radiology													
Respiratory													
Security													

3. Success Criteria

    a. **Key deliverables**—The success of the project will depend on executing detailed transition action plans that are coordinated with interdisciplinary and support departments; preparing workflow diagrams that clearly communicate processes and roles; maintaining risk and issue logs that track challenges to successful completion; preparing contingency plans that account for simultaneous operations and worst-case scenarios; and conducting realistic rehearsals prior to actual moves. The ultimate deliverable for this project is a validated instance of smooth communications during a public safety or medical emergency in the town.

    b. **Traceability to Business Requirements Document (BRD)**

Business Requirement	User Requirement	Project Deliverable
Provide state-of-the-art public safety and medical response services to the community with minimal disruption to on-going services.	1. Floor space layout that optimizes staff efficiency. 2. Preserve or enhance operations in new space. 3. Successful simultaneous operation of Emergency Communications Office at two locations for no more than one day.	1. Provide floor plans & detailed space layout for familiarization. 2. Find or construct process maps for AS IS and TO BE operations. 3. Plan and rehearse simultaneous operation of Emergency Communications Office in old and new space for up to one day.

    c. **Quality metrics**—Project team meetings will serve as the primary forum to gauge the quality and progress in each of the above deliverables.

4. Assumptions

    a. The dates currently shown in the Priorities section of this SOW are preliminary as of 10/2006. However, if they move, they will most likely do so in weekly increments.

    b. Participation from all elements of each department will be uniform.

5. Constraints

    a. Communication equipment configurations have been frozen and no further changes are open.

    b. Planning and execution of transition work is in addition to normal municipal operations and will require priority management and labor overtime for success.

6. Change Control Process

Any substantive changes to any part of this project will follow a formal, documented process that includes all stakeholders, to ensure proper participation and consensus. The following change request form will facilitate this process.

Project Change Request Form	
Project Name:	
Project Number:	
Change requested by (Name/Title):	
Date change requested:	
Description of requested change:	
Justification:	
Impact on scope:	
Impact on deliverables:	
Impact on schedule:	
Impact on staffing resources:	
Impact on financial resources:	
Impact on business processes and workflows:	
Impact on risk:	
Impact on quality:	
Other impact:	
Alternatives:	
Comments:	

Attachments

<<List any attachments below>>

Approval/Disapproval

Name	Title	Approved	Disapproved	Date
		/s/ attach e-mail		
		/s/ attach e-mail		

Final decision communicated to project team on _____ by _____

Date   Name

Attachments

A. **Business Requirements Document (Project Assessment Form)**

ATTACHMENT A TO SOW–PROJECT ASSESSMENT FORM

Date: 9-27-06	MUNICIPAL EMERGENCY COMMUNICATIONS SYSTEM	Approved by: Leadership Team
PROJECT DETAILS	**OBJECTIVES**	**ACTIONS**
WHAT? • problem/ oppty • needs • solutions • costs • out of scope • assumptions • constraints	1. Completed floor plans & detailed space layout for familiarization. 2. Staff comfort with & understanding of new space. 3. Re-evaluate all processes for preservation in new space. 4. Determine/confirm sequence of transitions. 5. Simultaneous operation of Emergency Communications Office at two locations for <1 day.	1. Acquire floor plans & detailed space layout. 2. Canvass core team and determine comfort criteria and level. 3. Find or construct process maps for AS IS and TO BE. 4. Review individual units' transition plans with unit managers. 5. Prepare plans for simultaneous operation of Emergency Communications Office in old and new space for up to one day.
WHEN? • crisis? • short-term • long-term • key milestones • underway?	1. Early move-in target is 1/23/08. 2. Weekly status meetings needed.	1. Determine lead time and availability of space for rehearsals. 2. Convene weekly core team meetings. 3. Confirm mutual understanding & details on communications changeover.

WHO?
• stakeholders (orgn & indiv)
• PMs
• core team
	1. Ensure that all support services are involved – e.g., facilities, telecom, transporters, interpreters, labs, etc.
	2. Ensure that HR oversees interaction with unions.
		1. Form core team
		  • both clinical and nonclinical leaders
		  • create Task-Team Matrix for interdisciplinary coordination
		  • list all emergency functions & confirm that they are included in planning & status mtgs.
		2. Confirm union and HR participation.

WHY?
• goal
• bus rqmts
• risks
• success=?
	1. Minimal impact on municipal, police, and hospital operations.	1. Practice sessions
		  • rehearsals in mock-up rooms & later in real space
		  • from patient and caregiver perspectives
		  • coordinate with emergency drill (volunteers)

HOW?
• approach
• tactics
• priorities
	1. Each unit has its own transition plans.
	2. Each unit has included contingency plans for worst-case scenarios.
		1. Coordinate across units
		  • Construct command center for overall project tracking.
		  • Integrate individual transition plans where there are inter-dependencies.
		2. Review worst-case contingency plans by unit for 3 priorities:
		  • red = life-threatening
		  • yellow = cost impact
		  • blue = patient impact (delay, confusion, etc.)

B. **Project Budget Details**—$1,695,000. See minutes of Town Hall Meeting from June 1, 2006.

C. **Project Plan**—Individual project plans provided by each of the three areas that are moving. Following is a project plan template.

EMERGENCY COMMUNICATIONS SYSTEM

Attachment C – Project Plan

Action No.	% Complete	Action Item	Assigned to	Date Due	Date Completed	Dependent on	COMMENTS
PRE-MOVE							
1.00		**Build Operations Teams**					
1.10		Emergency Preparedness Office					
1.20		Police Precint # 1					
1.30		Police Precint # 2					
1.40		Interdisciplinary & Support Svcs					
1.50		City Memorial Hospital					
1.60		County General Hospital					
1.70		University Medical Ctr					
1.80		Westside Medical Ctr					
1.90		St. Mary's Hospital					
2.00		**Conduct Planning Mtg**					
2.10		Send out invitations with agenda					
2.20		Hold meeting					
3.00		**Conduct Status Mtgs**					
3.10		Send out invitations with agenda					
3.20		Hold meeting					
4.00		**Conduct Rehearsal(s)**					
4.20		Send out invitations with agenda					
4.20		Hold meeting					
MOVE							
POST-MOVE							

D. **Risk Management**—All risks will be recorded in the Risk Log, assigned a priority level (red/yellow/green), and managed to completion. Red risks that pass an assigned due date will be subject to escalation in accordance with Attachment F below.

E. **Risk Log**—Templates are following. The official log is posted on the shared drive (see Documentation Protocol, Attachment H).

EMERGENCY COMMUNICATIONS SYSTEM

Proj. Mgr. Name
Contact phone no.

Attachment E in SOW -- Issues/Risk Management Log

DEPT	RISK No.	RISK DESCRIPTION	IDENTIFIED BY	DATE ENTERED IN THIS LOG	IMPACT on success	LIKELIHOOD of occurrence	ACTION PLAN [Assignee]	READY STATUS	PROGRESS as of mm/dd/yy
								R	= less than 50% READY
								Y	= 50-75% READY
								G	= greater than 75% READY

(Open issues/risks)

EMERGENCY COMMUNICATIONS SYSTEM

Proj. Mgr. Name
Contact phone no.

Attachment E in SOW - Issues/Risk Management Log

DEPT	RISK No	RISK DESCRIPTION	IDENTIFIED BY	DATE ENTERED IN THIS LOG	DATE CLOSED	CLOSED BY	REASON FOR CLOSURE [Decision-maker]

(Closed issues/risks)

F. **Escalation Policy—Emergency Communications System**

   Tasks, issues, and decisions that involve negotiable or debatable viewpoints identified as **red** in the Issues Log must reach resolution within **one week**, or they must automatically escalate to the next higher level of management for arbitration. Escalations will be documented via e-mail to all involved/affected parties. The final level of escalation will be James Kirkpatrick (Mayor).

G. **Communications Plan**—A comprehensive, detailed communications plan is available from The Emergency Preparedness Office.

H. **Documentation Protocol—Emergency Communications System**

   Shared storage space named **Emergency Comm Sys Upgrade** has been created on drive **TOWNPUBSAFETY01** with access limited to project team members. Folders on the drive include:

   • **Project and team organization**—to include the Statement of Work, Task-Team Matrix, and other project-level documents

   • **Transition plans and schedules**—detailed action plans from each department for operations during the pre-move, move, and post-move periods

   • **Meeting agendas and minutes**

   • **Workflow documents**—diagrams for both existing "AS IS" workflows in the current environment as well as planned "TO BE" workflows in the new environment

   • **Contingency plans**—plans that account for simultaneous operations and for worst-case scenarios

   • **Issues and risks**—including the project Issues Log

   • **Miscellaneous**—for documents that do not fit into any other folder categories

   • **Archives**—Use it to store early/obsolete versions of project files so that we can refer back to them if need be. All of the files in this folder have the prefix "Archive-" appended to the original file name, which makes it easy both to identify them and to keep them distinct from the current version which is posted in one of the other main folders.

I. **Test Strategy**—Rehearsals in mock facilities will be available in late-November/early December.

J. **Training Strategy**—Training on new equipment will be available and is being coordinated by the Municipal Engineering Department. The completed schedule will be posted as Attachment J in this statement of work.

Approvals

Name	Title	Function	Date
Cicely Tangery	Director, Municipal Emergency Preparedness	Municipal Coordinator	October 15, 2006
Sgt. Edward Jenkins	Emergency Officer	Coordinator, Precinct No. 1	October 15, 2006
Lt. Emily Harris	Emergency Officer	Coordinator, Precinct No. 1	October 15, 2006
Grayson Edwards	Assistant Administrator	Coordinator, City Memorial	October 15, 2006
Paul Jenkins	VP Operations	Coordinator, County General	October 15, 2006
Dr. Lauren Fields	Medical Director, Emergency Department	Coordinator, University Medical Center	October 15, 2006
Katherine Thompson	Director, Emergency Department	Coordinator, Westside Medical Center	October 15, 2006
Sister Mary Pancratia	Senior Director, Emergency Department	Coordinator, St Mary's Hoppital	October 15, 2006
Michael Zambruski	Director, Project Management Office	Advisor	October 15, 2006

Appendix 12

Project Roster Template

PROJECT ROSTER

Name	Title	Department (or Firm)	Area of Specialty	Role/ responsibility	Phone contact	Alternate	Alternate's phone	Admin. Assistant

Appendix 13

Project Plan Templates

<< PROJECT NAME >>

ID	ⓘ	% Complete	Task Name	Duration	Start	Finish	Predecessors	Resource Names	3rd Quarter / 4th Quarter / 1st Quarter / 2nd Quarter
1		0%	**1 INITIATE**	**10 days**	**Thu 8/3/06**	**Wed 8/16/06**			
2		0%	1.1 Develop project Charter	5 days	Thu 8/3/06	Wed 8/9/06			
3		0%	1.2 Obtain Charter approval	5 days	Thu 8/10/06	Wed 8/16/06	2		
4									
5		0%	**2 PLAN**	**70 days**	**Thu 8/17/06**	**Wed 11/22/06**			
6		0%	2.1 Prepare Requirements Work Plan	10 days	Thu 8/17/06	Wed 8/30/06	3		
7		0%	2.2 Prepare Business Requirements Document	30 days	Thu 8/31/06	Wed 10/11/06	6		
8		0%	2.3 Prepare Statement of Work	30 days	Thu 10/12/06	Wed 11/22/06	7		
9									
10		0%	**3 EXECUTE**	**115 days**	**Thu 11/23/06**	**Wed 5/2/07**			
11		0%	3.1 Develop specifications and solutions	15 days	Thu 11/23/06	Wed 12/13/06	8		
12		0%	3.2 Create & test preliminary solution	10 days	Thu 12/14/06	Wed 12/27/06	11		
13		0%	3.3 Plan testing for final solution	5 days	Thu 12/14/06	Wed 12/20/06	11		
14		0%	3.4 Develop final solution	20 days	Thu 12/28/06	Wed 1/24/07	12		
15		0%	**3.5 Test final solution**	**10 days**	**Thu 1/25/07**	**Wed 2/7/07**			
16		0%	3.5.1 against requirements	10 days	Thu 1/25/07	Wed 2/7/07	14		
17		0%	3.5.2 for usability	10 days	Thu 1/25/07	Wed 2/7/07	14		
18		0%	3.5.3 for integration across departments	10 days	Thu 1/25/07	Wed 2/7/07	14		
19		0%	3.6 Finalize solution documentation	30 days	Thu 2/8/07	Wed 3/21/07	15		
20		0%	3.7 Develop training materials and schedule	30 days	Thu 2/8/07	Wed 3/21/07	15		
21		0%	3.8 Plan for transition into production	40 days	Thu 1/25/07	Wed 3/21/07	14		
22		0%	3.9 Transition to production	0 days	Wed 3/21/07	Wed 3/21/07	21		◆ 3/21
23		0%	3.10 Post-transition warranty support to production	30 days	Thu 3/22/07	Wed 5/2/07	22		
24									
25		0%	**4 CONTROL**	**114 days**	**Wed 11/22/06**	**Tue 5/1/07**			
26		0%	4.1 Conduct project kickoff meeting	0 days	Wed 11/22/06	Wed 11/22/06	8		◆ 11/22
27	↻	0%	**4.2 Weekly project team meetings**	**114 days**	**Thu 11/23/06**	**Tue 5/1/07**	26		
67	↻	0%	**4.3 Monthly stakeholder meetings**	**101 days**	**Thu 11/23/06**	**Thu 4/12/07**	26		
77									
78		0%	**5 CLOSE**	**6 days**	**Thu 5/3/07**	**Thu 5/10/07**			
79		0%	5.1 Conduct project debriefing	3 days	Thu 5/3/07	Mon 5/7/07	23		
80		0%	5.2 Document and distribute lessons learned	3 days	Tue 5/8/07	Thu 5/10/07	79		

Appendix 13.1   (Microsoft Project)

<< PROJECT NAME >>

Project Manager: _____
Telephone: _____

Taks #	% Complete	Task Name	Duration	Start	Finish	Prececessors	Resources
1	0%	**INITIATE**	10 days	08/03/06	08/16/06		
2	0%	Develop project Charter	5 days	08/03/06	08/09/06		
3	0%	Obtain Charter approval	5 days	08/10/06	08/16/06	2	
4							
5	0%	**PLAN**	70 days	08/17/06	11/22/06		
6	0%	Prepare Requirements Work Plan	10 days	08/17/06	08/30/06	3	
7	0%	Prepare Business Requirements Document	30 days	08/31/06	10/11/06	6	
8	0%	Prepare Statement of Work	30 days	10/12/06	11/22/06	7	
9							
10	0%	**EXECUTE**	115 days	11/23/06	05/02/07		
11	0%	Develop specifications and solutions	15 days	11/23/06	12/13/06	8	
12	0%	Create & test preliminary solution	10 days	12/14/06	12/27/06	11	
13	0%	Plan testing for final solution	5 days	12/14/06	12/20/06	11	
14	0%	Develop final solution	20 days	12/28/06	01/24/07	12	
15	0%	Test final solution	10 days	01/25/07	02/07/07		
16	0%	against requirements	10 days	01/25/07	02/07/07	14	
17	0%	for usability	10 days	01/25/07	02/07/07	14	
18	0%	for integration across departments	10 days	01/25/07	02/07/07	14	
19	0%	Finalize solution documentation	30 days	02/08/07	03/21/07	15	
20	0%	Develop training materials and schedule	30 days	02/08/07	03/21/07	15	
21	0%	Plan for transition into production	40 days	01/25/07	03/21/07	14	
22	0%	Transition to production	0 days	03/21/07	03/21/07	21	
23	0%	Post-transition warranty support to production	30 days	03/22/07	05/02/07	22	
24							
25	0%	**CONTROL**	114 days	11/22/06	05/01/07		
26	0%	Conduct project kickoff meeting	0 days	11/22/06	11/22/06	8	
27	0%	Weekly project team meetings	114 days	11/23/06	05/01/07	26	
28	0%	Monthly stakeholder meetings	101 days	11/23/06	04/12/07	26	
29							
30	0%	**CLOSE**	6 days	05/03/07	05/10/07		
31	0%	Conduct project debriefing	3 days	05/03/07	05/07/07	23	
32	0%	Document and distribute lessons learned	3 days	05/08/07	05/10/07	28	

Appendix 13.2 (Microsoft Excel)

EMERGENCY COMMUNICATIONS SYSTEM

Project Manager: Michael S. Zambruski
Telephone: 123-456-7890

Project Plan Template

Action No.	% Complete	Action Item	Assigned to	Date Due	Date Completed	Dependent on	COMMENTS
PRE-MOVE							
1.00		**Build Operations Teams**					
1.10		Emergency Preparedness Office					
1.20		Police Precint # 1					
1.30		Police Precint # 2					
1.40		Interdisciplinary & Support Svcs					
1.50		City Memorial Hospital					
1.60		County General Hospital					
1.70		University Medical Ctr					
1.80		Westside Medical Ctr					
1.90		St. Mary's Hospital					
2.00		**Conduct Planning Mtg**					
2.10		Send out invitations with agenda					
2.20		hold meeting					
3.00		**Conduct Status Mtgs**					
3.10		Send out invitations with agenda					
3.20		hold meetings					
4.00		**Conduct Rehearsal(s)**					
4.20		Send out invitations with agenda					
4.20		hold meeting					
MOVE							
5.00		Prepae script for simultaneous opn					
6.00		Operation at 2 sites simultaneously					
POST-MOVE							
7.00		Conduct post-move debriefing					
8.00		Document & distribute lessons learned					

Appendix 13.3   (Microsoft Excel)

Appendix 14

Completed Project Plan

PROJECT PLAN: Program Management Office Implementation

Michael S. Zambruski
123-456-7890

ID	% Complete	Task Name	Start	Finish	Duration	Predecessors	Resource
1	74%	1 INITIATE	Mon 7/16/07	Mon 8/6/07	16 days?		
2	82%	1.1 Develop Charter for PMO creation	Mon 7/16/07	Fri 7/27/07	10 days		Mike
3	90%	1.1.1 use template in handbook	Mon 7/16/07	Wed 7/25/07	8 days		
4	75%	1.1.2 define PMO structure and staffing	Mon 7/16/07	Fri 7/27/07	10 days		
5	50%	1.2 Obtain Charter signoffs	Mon 7/30/07	Mon 8/6/07	6 days?	2	
6	7%	2 PLAN	Tue 8/7/07	Mon 8/27/07	15 days?		Mike
7	35%	2.1 Reqmts Work Plan (RWP)	Tue 8/7/07	Thu 8/9/07	3 days?	5	
8	50%	2.1.1 Prepare draft RWP using handbo??	Tue 8/7/07	Wed 8/8/07	2 days	5	
9	5%	2.1.2 Obtain signoff on RWP & begin??	Thu 8/9/07	Thu 8/9/07	1 day?	8	
10	1%	2.2 Bus Rqmts Doc (BRD)	Thu 8/9/07	Mon 8/20/07	8 days?		
11	5%	2.2.1 Prepare draft BRD	Thu 8/9/07	Fri 8/9/07	1 day?	9FF	
12	0%	2.2.2 Prepare script for interviews	Fri 8/10/07	Fri 8/10/07	1 day?	11	
13	0%	2.2.3 Schedule interviews (1:1 or 1:man??	Mon 8/13/07	Mon 8/13/07	1 day?	12	
14	0%	2.2.4 Conduct interviews, meetings	Tue 8/14/07	Thu 8/16/07	3 days	13	
15	0%	2.2.5 Synthesize information into update??	Fri 8/17/07	Fri 8/17/07	1 day?	14	
16	0%	2.2.6 Obtain signoff on BRD	Mon 8/20/07	Mon 8/20/07	1 day?	15	
17	0%	2.3 Statement of Work (SOW)	Tue 8/21/07	Mon 8/27/07	5 days?	16	
18	2%	3 EXECUTE	Mon 7/16/07	Wed 11/21/07	93 days?		Mike,Bob
19	0%	3.1 Objective 1 - establish enterprise PMO	Tue 8/28/07	Mon 11/19/07	60 days		
20	0%	3.1.1 Annouce enterprise-wide	Tue 8/28/07	Mon 9/3/07	5 days	17	
21	0%	3.1.2 Formalize certification rqmts for B??	Tue 8/28/07	Mon 9/3/07	5 days	17	
22	0%	3.1.3 Complete BA & PM staffing	Tue 8/28/07	Mon 11/19/07	60 days	17	
23	2%	3.2 Objective 2 - integrate Bus Analysis+Proj Mgt	Mon 7/16/07	Wed 11/21/07	93 days		
24	8%	3.2.1 Develop prioritization criteria	Mon 7/16/07	Fri 8/31/07	35 days		
25	0%	3.2.2 Assign priorities to candidate proj??	Mon 8/13/07	Fri 11/16/07	70 days		
26	0%	3.2.3 Assign projects to BAs & PMs	Tue 8/28/07	Wed 11/21/07	62 days	22SS,28	
27	7%	3.3 Objective 3 - implement PMO guidelines	Mon 7/16/07	Wed 9/5/07	38 days		
28	10%	3.3.1 consolidate into one handbook	Mon 7/16/07	Wed 8/22/07	28 days	22SS	
29	0%	3.3.2 post on website(s)	Thu 8/23/07	Wed 9/5/07	10 days	28	
30	0%	3.4 Objective 4 - develop training curricula & sched	Tue 8/28/07	Fri 11/2/07	49 days?		
31	0%	3.4.1 work with BAs & PMs for best fit	Tue 8/28/07	Mon 9/10/07	10 days	22SS	
32	0%	3.4.2 work with ESI on time, $, instruct??	Mon 9/3/07	Fri 10/5/07	25 days	31SS	
33	0%	3.4.3 finalize courses & schedules	Mon 10/8/07	Fri 10/19/07	10 days	31,32	
34	0%	3.4.4 coordinate with HR	Mon 10/8/07	Fri 10/19/07	10 days	31,32	
35	0%	3.4.5 attach to Performance Evals	Mon 10/22/07	Fri 11/2/07	10 days?	34	
36	0%	4 CONTROL	Wed 10/3/07	Wed 12/19/07	56 days		Mike,Team
37	0%	4.1 BA training	Mon 10/22/07	Fri 12/14/07	40 days	33	
38	0%	4.2 PM training	Mon 10/22/07	Fri 12/14/07	40 days	33	
39	0%	4.3 Project Mgmt Audits	Thu 11/22/07	Wed 12/19/07	20 days	26	
40	0%	4.4 Business Analysis Forums	Wed 10/3/07	Wed 12/5/07	46 days		
44	0%	4.5 Project Mgmt Forums	Fri 10/5/07	Fri 12/7/07	46 days		
48	0%	5 CLOSE	Thu 12/20/07	Wed 1/30/08	30 days	36	Team
49	0%	5.1 Operationalization	Thu 12/20/07	Wed 1/30/08	30 days		
50	0%	5.2 Lessons learned	Thu 12/20/07	Wed 12/26/07	5 days		

Timeline (Gantt chart) across: 3rd Quarter (Jun, Jul, Aug, Sep), 4th Quarter (Oct, Nov, Dec), 1st Quarter (Jan, Feb)

Appendix 15

Issues/Risk Management Plan Template

<<Project Name>>

ATTACHMENT D to the Statement of Work

Issues/Risk Management Plan

1. Any elements of conflict or uncertainty that can impact project success will be identified and proactively managed for optimal results. The primary mechanism for this process will be the Issues/Risk Log (Attachment E to the standard Statement of Work).
2. The Issues/Risk Log will provide the following information at a minimum for all OPEN items:
   a. organizational unit(s) affected
   b. unique number assigned to each item in the log
   c. full item description
   d. person identifying the item
   e. date entered into the log
   f. estimated *impact* on the project scope, schedule, budget, or deliverables
   g. estimated *likelihood* or probability of occurrence
   h. preliminary action needed, together with person assigned to manage the action
   i. ready status (red/yellow/green)
   j. regularly scheduled summaries of progress to date (e.g., weekly, monthly, etc.)
3. The Issues/Risk Log will provide the following information at a minimum for all CLOSED items:
   a. organizational unit(s) affected
   b. original number assigned to the item being closed
   c. full item description
   d. person who originally identified the item
   e. date originally entered into the OPEN log
   f. date closed
   g. person closing the item
   h. reason for closure
4. The entire Issues/Risk Log will be updated at least once per week and will constitute a standing agenda item at all project team meetings, regardless of their frequency.
5. No-fault escalation to the next higher level of management will occur when any issues or risks remain open for longer than the maximum period allowable for the project, as specified in the **Escalation Policy** (Attachment F to the standard Statement of Work).

Issues/Risk Management Plan Template

<<Project Name>>

ATTACHMENT D to the Statement of Work

Issues/Risk Management Plan

1. Any element of conflict or uncertainty that can impact project success will be identified and proactively managed for optimal results. The primary mechanism for this process will be the Issues/Risk Log (Attachment 6 to the standard Statement of Work).

2. The Issues/Risk Log will provide the following information, at a minimum for all OPEN items:
   a. organizational unit(s) affected
   b. unique number assigned to each item in the log
   c. full item description
   d. person identifying the item
   e. date entered into the log
   f. estimated impact on the project scope, schedule, budget, or deliverables
   g. estimated likelihood or probability of occurrence
   h. preliminary action needed, together with person assigned to manage the action
   i. ready status (active/green)
   j. regular, scheduled summaries of progress to date (e.g., weekly, monthly, etc.)

3. The Issues/Risk Log will provide the following information at a minimum for all CLOSED items:
   a. appropriate unit number affected
   b. original number assigned to the item being closed
   c. full item description
   d. person who originally identified the item
   e. date originally entered into the OPEN log
   f. date closed
   g. person closing the item
   h. reason for closure

4. The entire Issues/Risk Log will be updated at least once per week and will be distributed to the project team...

Appendix 16

Issues and Risk Log Template

Attachment E in SOW — Issues/Risk Management Log

DEPT	RISK No.	RISK DESCRIPTION	IDENTIFIED BY	DATE ENTERED IN THIS LOG	IMPACT on success	LIKELIHOOD of occurrence	ACTION PLAN [Assignee]	READY STATUS	PROGRESS as of mm/dd/yy	
								R	= less than 50% READY	
								Y	= 50-75% READY	
								G	= greater than 75% READY	

Appendix 16.1   (Open risks)

Attachment E in SOW - Issues/Risk Management Log

DEPT	RISK No.	RISK DESCRIPTION	IDENTIFIED BY	DATE ENTERED IN THIS LOG	DATE CLOSED	CLOSED BY	REASON FOR CLOSURE [Decision-maker]

Appendix 16.2   (Closed risks)

Appendix 17

Completed Issues and Risk Log

EMERGENCY COMMUNICATIONS PROJECT

Michael S. Zumbruski
123-334-2438

ISSUES/RISK LOG

DEPT	RISK No.	RISK DESCRIPTION	IDENTIFIED BY	DATE ENTERED IN THIS LOG	IMPACT on successful survey	LIKELIHOOD of inclusion in survey	ACTION PLAN [Assignee]	READY STATUS	PROGRESS as of 11/21/2006
								R	less than 50% READY
								Y	50-75% READY
								G	greater than 75% READY
Police Precinct #1	1	Processes are not being followed	Officer Taylor	11/14/2006	High	High	To be covered in training push. Must track progress weekly. [Victoria]	Y	11-21-06 [Victoria] Written procedures are now available, and rehearsals are being planned.
St. Mary's	2	Education delivery	Chip Williams	11/21/2006	Med	Med	To be covered in training push. Must track progress weekly. [Sr. Pancratia]	Y	
Municipal Human Resources	3	Orientation and readiness of new employees (contract)	Donna Swanson	11/21/2006	Low	Low	Facilities and HR working on plan to prepare self-directed orientation for new hires. [Cicely]	Y	11/19/2006 [Karen] Ready by 12/15/06.

Appendix 17.1   (Open risks)

EMERGENCY COMMUNICATIONS PROJECT

Michael S. Zumbruski
123-334-2438

ISSUES/RISK LOG

DEPT	RISK NUMBER	RISK DESCRIPTION	IDENTIFIED BY	DATE ENTERED IN THIS LOG	DATE CLOSED	CLOSED BY	REASON FOR CLOSURE [Decision-maker]
Cross-functional	4	Infrastructure support — turnaround time on contract and policy approvals is taking too long.	Mike Zambruski	12/10/2006	12/15/2006	Mike Zambruski	Joint subcommittee was created to accelerate infrastructure issues. [Victor, VP]

Appendix 17.2   (Closed risks)

Test Planning Template

Pre-implementation Testing and Validation Plan

Business Procedure: _____

Action Steps	Step Complete		Comments
	Date	**Y/N**	
Procedure complete and validated			
Training plan developed			
Staff training completed (see validation roster below)			
Staff compliance observed (see validation roster below)			
Implementation go-live date communicated to staff			

Staff Testing and Validation Roster						
Staff Member	Date	Procedure properly verbalized	Procedure properly performed	Compliant	Noncompliant	Scheduled date for review and revalidation

Appendix 19

Training Plan

Training Coordinator:_____
Telephone:_____

Organization	Telecommunications Procedures	Emergency Routes	Hospital Emergency Dept. Protocols	COMMENTS
Emergency Preparedness Office	10/16/2007	11/22/2007	12/12/2007	Both office and field staff will participate.
Police Precint # 1	10/16/2007	11/23/2007	12/13/2007	Try to avoid overtime.
Police Precint # 2	10/16/2007	11/24/2007	12/14/2007	Try to avoid overtime.
Interdisciplinary & Support Svcs	10/16/2007	11/25/2007	12/15/2007	Ensure all disciplines participate.
City Memorial Hospital	10/17/2007	11/26/2007	12/16/2007	Highlight differences between hospitals.
County General Hospital	10/18/2007	11/27/2007	12/17/2007	Highlight differences between hospitals.
University Medical Ctr	10/19/2007	11/28/2007	12/18/2007	Highlight differences between hospitals.
Westside Medical Ctr	10/20/2007	11/29/2007	12/19/2007	Highlight differences between hospitals.
St. Mary's Hospital	10/21/2007	11/30/2007	12/20/2007	Highlight differences between hospitals.

Appendix 20

Change Request Form Template

Project Change Request Form

Project Name:	
Project Number:	
Change requested by (Name/Title):	
Date change requested:	
Description of requested change:	
Justification:	
Impact on scope:	
Impact on deliverables:	
Impact on schedule:	
Impact on staffing resources:	
Impact on financial resources:	
Impact on business processes & workflows:	
Impact on risk:	
Impact on quality:	
Other impact:	
Alternatives:	
Comments:	

Attachments

<<List any attachments below>>

Approval/Disapproval

Name	Title	Approved	Disapproved	Date
		/s/ attach e-mail		
		/s/ attach e-mail		
		/s/ attach e-mail		
		/s/ attach e-mail		

Final decision communicated to project team on _____ **by** _____
                                            Date   **Name**

Appendix 21

Escalation Policy Template

High-priority tasks, issues, and decisions must reach resolution by their assigned target dates, or they must automatically escalate to **<<name>>**, the next higher level of management, for arbitration. This is especially necessary when negotiable or debatable viewpoints are involved that can delay timely resolution. As a normal practice, all items marked as RED in the Issues/Risk Log are subject to this policy.

Such escalations will be documented at least via e-mail to all involved/affected parties. Such escalation will continue up the management hierarchy until a final decision is reached. Accordingly, each member of the executive team must be made aware of his or her potential role in this process.

This no-fault Escalation Policy is included as Attachment F to the project Statement of Work. It is ultimately the responsibility of **<<name>>**, the project manager, to ensure that it is followed.

Appendix 27

Escalation Policy Template

High-priority tasks, issues, and decisions must reach resolution by their assigned target dates, or they must automatically escalate to <<name>>, the next higher level of management for arbitration. This is especially necessary when negotiable or debatable viewpoints are involved that can delay timely resolution. As a general practice, all items marked as KEY in the Issue/Risk Log are subject to this policy.

Such escalations will be documented at least via e-mail to all involved/affected parties. Such escalation will continue up the management hierarchy until a final decision is reached. Accordingly, each member of the executive team must be made aware of his or her potential role in this process.

This no-fault Escalation Policy is included as Attachment F to the project Statement of Work. It is ultimately the responsibility of <<name>>, the project manager to ensure that it is followed.

Appendix 22

Communications Plan Template

1. **Format for communications**. The project manager will distribute guidelines and templates for use in all official project communications, both internal and external.
2. **Media**. All official project communications will be distributed in written form, either in the text body of an e-mail or as an e-mail attachment. Copies will be retained in the project repository at Web site (insert URL of Web site). Voice messages will always be supported with e-mails.
3. **Points of control**. The project manager or his or her designee will coordinate all communication that involves the entire project team and any external partners. Team leaders will ensure that communication within and between teams follows this Communications Plan.
4. **Kick-off meeting**. The project manager will arrange a project kick-off session involving all stakeholders to review in detail the Statement of Work as soon as it has been completed.
5. **Biweekly project team sessions**. Every other week, at a time to be determined, the entire project team identified in paragraph 2 of the Statement of Work will meet to review the following items at a minimum:
   a. overall status of individual action plans
   b. completeness of documentation
   c. risk management
   d. outstanding escalations
   e. status of testing and training, as needed or applicable.
6. **Weekly status sessions**. Each week, on a day to be determined, the project team leaders will meet at least by phone for 30 minutes to report on the following:
   a. progress of assigned action items currently due, according to the project plan
   b. new risks
   c. new escalations.
7. **Meeting minutes**. The project manager (or alternate) will post minutes from all project-level meetings in the Web-based repository identified for this project. See SOW Attachment H, Documentation Protocol. Team leaders will be responsible for posting minutes to meetings that they conduct.

Meeting Agenda Template and Sample Meeting Agenda

Meeting Agenda Template

MEETING AGENDA			
Project:			
Purpose:			
Date: Time: Place:			
<u>Invitees</u>			
Time	**Agenda Item**		**Lead**
NOTES			

Sample Meeting Agenda

MEETING AGENDA		

Project: **National Security Number (NSN) Project**
Purpose: **Kick-Off Meeting**

Date: 7/20/06
Time: 8:00 am – 10:00 am
Place: Conference Room 123

Invitees: Debbie Flora, Ann Garner, Nancy Horla, Andy Jacobs, Gary Rumento, Carol Mason, Lauren O'Brien, Josh Laggy, Karen Randolph, Sharon Stone, Kathleen O'Hara, Mike Zambruski

Time	Agenda Item	Lead
8:00	Introductions and Meeting Purpose	M. Zambruski
8:15	NSN General Overview	L. O'Brien
8:25	NSN Project • Project Charter and Approval • Team Roster and Alternates • Access to Project Documents on Web Repository	L. O'Brien, A. Jacobs
8:45	NSN Business Requirements Document (BRD)—quick review	L. O'Brien
9:15	NSN Statement of Work (SOW)—detailed review	A. Jacobs
9:45	Next Steps	All
10:00	Meeting ends	
NOTES		

Appendix 24

Meeting Minutes Template and Sample Meeting Minutes

Meeting Minutes

Organization:

Project Name:

Team or Committee Name:

Date and Time of Meeting:

Meeting Location (building, room, etc.):

Team Members Attending:	
Team Members Absent:	
Excused:	Anyone who is not on the team but who was invited and could not attend.
Visitors:	This includes anyone who is not on the team but who was invited, plus walk-ins who were not originally invited.

Agenda Item	Discussion	Action/Follow-up (assignee & due date)

Sample Meeting Minutes

Meeting Minutes

Enterprises, Inc.
Transformation Project
Project Core Team
July 6, 2006 at 0930
Conference Room 1

Team Members Attending:	Emmett Walsh, Gary Sensei, Karen Phillips, John Smith, Bob Wang, Mike Zambruski
Team Members Absent:	Stacy Peters
Excused:	George Escobar
Visitors:	Wendy Jersey

Agenda Item	Discussion	Action/Follow-up(assignee & due date)
Process flows ("As Is")	• Services—weekly meetings are still occurring, progress is steady, 4 service categories are completely mapped in an Excel table; next will be flow diagrams • Capital—progress is steady, 40 flows have been mapped, additional 16 pertain to members' hospital processes • Supplies—processes mapping is being completed in time for phase 1 implementation (8/14/06)	• None
Project teams' Statements of Work	• Services and Capital working teams will modify their SOWs to state that current target dates will be delayed due to redeployment of DBMS expertise to DBMS implementation. However, precise amount of delay is not yet certain, since it depends on when DBMS resources are again available to the Capital and Services working teams. • Key Performance Indicators (KPIs) presented at 7/3/06 Exec. Team Mtg. will be incorporated into SOWs as deliverable metrics to the extent that they apply to the current scope of work. In those cases where applicability of KPI is uncertain for the Program Transformation, or where incorporating them impacts the existing scope of work, guidance from the Exec. Team will be sought.	• Subteam leaders (John, Karen, Gary, George, Emmett) will modify dates by next meeting (7/13/06).

Appendix 25

Documentation Protocol Template

- **Archives**—Used to store early or obsolete versions of project files so that we can refer back to them if need be. All the files in this folder have the prefix "Archive-" appended to the original file name, which makes it easy both to identify them and to keep them distinct from the current version which is posted in one of the other main folders (below).
- **Budget documents**—funds, staff, equipment, supplies.
- **Design documents**—conceptual, preliminary, detailed, and final designs.
- **Issues and Risks Log.**
- **Meeting agendas.**
- **Meeting minutes.**
- **Miscellaneous**—anything not identified elsewhere in this protocol (e.g., procurement plans).
- **Project definition documents.**
  - Project Business Case
  - Project Charter
  - Requirements Work Plan (RWP)
  - Business Requirements Document (BRD)
  - Statement of Work (SOW)
- **Project planning and control documents**
  - Project Plans
  - Contingency Plans
  - Escalation Policy
  - Communications Plan
  - Documentation Protocol
  - Project Audits
- **Reference documents**—market studies, quality data, industry information, etc.
- **Training and education documents.**
- **Validation and testing documents**—test plans and validation records.
- **Workflow documents**—diagrams and charts depicting current and proposed workflows and process flows (both logical and physical).

Appendix 26

Lessons Learned Template

POST-PROJECT LESSONS LEARNED

PROJECT NAME:	PROJECT MGR:	DATE PREPARED:
Project Start Date:	Original Project End Date:	Actual Project End Date:

WHAT CONTRIBUTED TO SUCCESS?

1.

WHAT HINDERED SUCCESS?

1.

PROJECT CHARACTERISTICS

Was the project **planned** properly?	
Were **users** involved in planning?	
Were **risks** identified & managed?	
Were **contingency plans** developed?	
Was the **decision structure** clear?	
Was **communication** timely?	

LESSONS LEARNED

What could have been done differently?	
Why wasn't it done?	
Where will these Lessons Learned be stored for retrieval by others?	

Prepared
by: _____

Date: _____

Index